# MR. TOMPKINS
# GETS SERIOUS

## The Essential George Gamow

Edited by Robert Oerter

Foreword by R. Igor Gamow

*The* MASTERPIECE SCIENCE *Edition*

Pi Press
New York

PI PRESS

An imprint of Pearson Education, Inc.

1185 Avenue of the Americas, New York, New York 10036

© 2006 by Pearson Education

Pi Press offers discounts for bulk purchases. For more information, please contact U.S. Corporate and Government Sales, 1-800-382-3419, corpsales@pearsontechgroup.com. For sales outside the U.S., please contact International Sales at international@pearsoned.com.

Printed in the United States of America

First Printing

Library of Congress Cataloging-in-Publication Data

Gamow, George, 1904-1968.
    Mr. Tompkins gets serious : the essential George Gamow /
George Gamow.-- The masterpiece science ed.
    p. cm.
    ISBN 0-13-187291-5
    1. Physics. 2. Matter--Properties. I. Title.
    QC71.G26 2006
    530--dc22
                                    2005025070

Pi Press books are listed at www.pipress.net.

Pearson Education LTD.
Pearson Education Australia PTY, Limited.
Pearson Education Singapore, Pte. Ltd.
Pearson Education North Asia, Ltd.
Pearson Education Canada, Ltd.
Pearson Educatión de Mexico, S.A. de C.V.
Pearson Education—Japan
Pearson Education Malaysia, Pte. Ltd.

# Contents

# Foreword
## *Memories of My Father*

I've heard that Scott Carpenter was the last astronaut to fly a human spacecraft "by the seat of his pants." I believe that Father was one of the last great scientists to do science by the seat of his pants. Big science today, and particularly big physics, is done with large groups of people using magnificent and expensive machines. Nobody flies by the seat of their pants anymore.

But Father was first and foremost a storyteller. His scientific stories have been shared for more than a generation, including the one about the "Big Bang of the Universe," but I would like to share some stories that he told and retold—ones that reveal another, lighter side.

### *Banned In Boston!*

I have been asked why I enjoy being so controversial. My typical response is that it is probably genetic.

In 1958, Father published his first popular undergraduate physics textbook , *Matter, Earth, and Sky,* which the material in this volume is taken from. I have the first printing of the first edition of this book on my shelf—even though my German shepherd Kim chewed it up. It contains a picture of a famous work of art not found in other printings of the book.

*Matter, Earth, and Sky* was banned in Boston to my father's delight. One of the universities in Boston had ordered about 300 copies of the book. While the potential buyers were paging through it they came to page 521, where Father discusses the origin of our Milky Way. But instead of using a real photograph of the Milky Way, which of course we can all see just by looking up at the night sky, Father wanted a painting by Tintoretto showing little Hercules sucking on the nipple of Hera. Hercules sucked so hard that the milk from her nipple spread throughout the sky and formed the Milky Way. The caption read, "Figure 20-15, *The Origin of the Milky Way* by Tintoretto, representing a Greek Cosmological Myth. The baby Hercules was brought in to be fed by Hera the Goddess of home life. Being very strong even at this tender age, Hercules applies too much pressure spraying milk out of both nipples and forming the stars of the Milky Way. The meaning of the eagle and the two peacocks is not clear." The powers that be at the university took one look at this and said, "No." Prentice Hall lost the sale of the books. In the second edition, they substituted a very unimaginative but accurate telescopic view of the Milky Way.

So, along with *Lady Chatterley's Lover*, and James Joyce's *Ulysses*, Father was pleased to report that his book had been banned in Boston.

## Father Makes Playboy

One day one of Father's physics students told him that he was in *Playboy*, he went down to the local newsstand and bought 25 copies!

*Playboy* used to have a column called "The Playboy Advisor," written for young men who presumably could ask the advisor personal questions, usually of a sexual nature. A young man wrote that he and another couple had gone to a mountain lodge for a day of skiing and dinner. One student couple was from a liberal arts background, and the other couple had a physics background. After a day of skiing they headed to a ski lodge for some warm slog in front of a warm fire. One of the liberal-arts students said to his girlfriend, "Oh be a fine girl kiss me right now sweetheart." The science students laughed out loud. Embarrassed, the liberal-arts student said, "What are you laughing about?" The physics couple replied, "If you knew a little bit more physics you would know."

So the embarrassed student asked the *Playboy* Advisor why the other couple was laughing. The Advisor explained: in the old days when there weren't very good telescopes you saw some stars were bright and some not so bright. Early astronomers made an a, b, c, d, e, f, etc., scale of brightness assuming that the stars that were the brightest were the closest and the stars that were the dimmest were the furthest. Later research determined that a star could appear quite dim because it was absolutely dim and not necessarily far away. So, the luminosity scale had to be reorganized. The absolute brightness of a star is called the luminosity of the star—more formally known as the stellar spectroscopic magnitude. Since the star's absolute distances are now known, the a, b, c, d, etc., scale changed to o, b, a, f, g, k, m, r, n, s. Father used the mnemonic "Oh Be A Fine Girl Kiss Me Right Now Sweetheart" when he was teaching astronomy. I'm not sure who originally devised the mnemonic but the *Playboy* Advisor informed the student that a professor at the University of Colorado, Professor George Gamow, had used the mnemonic when teaching an astronomy class. Knowing him, it is a wonder he bought only 25 copies.

## The Golden Crown Experiment Goes Awry

I started at the University of Colorado in 1958 and registered as a freshman to take my father's new course called "Matter, Earth, and Sky." He taught this undergraduate physics course for non-majors using the textbook by the same name. This was the first time I had been in a classroom with Father. There were about a hundred students in that course and I bet that all of them remember it.

Father was convinced that the better you are as a theoretical physicist the worse you are as an applied physicist. He loved to tell a story about a huge explosion in Germany that leveled an entire factory. Nobody could figure out why the factory exploded. Finally somebody discovered that Wolfgang Pauli—who was probably the world's greatest theoretical physicist at the time—was in a train passing through the city when the factory exploded. Thus, the reason for the accident was attributed to the "Pauli Effect"—the great theoretical presence of Pauli caused a momentary deficit in practical application, things went awry, and the factory went up in flames.

In the freshman course, Father decided for the first time to demonstrate his own experiments to the class. He was an uninspired experimentalist when he was a graduate student at the University of Leningrad. Perhaps he thought he would give experimenting another try. He wanted to show the Archimedes experiment. Allegedly, Archimedes was asked to determine whether a gold crown that the king had worn was really pure gold. Archimedes couldn't scratch it or deface it, so he submerged it in water and measured the volume of water displaced, which allowed him to determine its density. He found that the crown was, in fact, pure gold.

The University of Colorado, of course, didn't make a real gold crown for the demonstration. The CU machine shop constructed one out of bronze instead. Professor Gamow came in to the lecture with his "gold" crown and explained how he was going to measure the volume of water displaced when he immersed the crown into a beaker of water. So far, so good.

He had a ring stand with the crown hung by a string and a 5000 cc glass beaker full of water beneath the crown. His idea was pretty simple—he would lower the crown into the beaker of water and measure how much water was displaced. He explained the principle and then he said, "Now I will lower the crown into the water." He turned the thumb screw, and the crown came crashing down. It smashed the beaker—glass shards and water went everywhere.

He stood wet from the waist down. The students in the front row got wet. They, of course, were laughing. Father meekly said, "Well, this is an experiment in dynamics, not in density." Father was always very persistent. He wondered what he could do to save his experiment. All the water was gone. But there was a sink built into the lecture table. He triumphantly announced, "The day is saved. We will submerge it into the water in the sink." He reached down and turned the faucet. It turned out to be steam instead of the water. A cloud of steam rose from the basin. His glasses were now completely fogged up. His hands frantically searched for the faucet without luck. Finally he got a hold of the nozzle and turned it off.

A girl next to me laughed so hard that she started to heave. I sat in the front row trying to make myself as small as possible. People were stomping. There was a tremendous amount of commotion. Father came out of the steam, and said, "Oh, wrong faucet." He looked down again, and said, "Ah, water." He turned on the water. But the water faucet had a long rubber hose attached. The hose came out of the sink and sprayed water everywhere. Father tried furiously to catch the hose. The students yelled, "Turn off the water! Turn off the water!" By this time the physics department secretaries and people out in the hall came pouring in. Father's assistant, Dalton, came to the rescue. He took Father by the elbows, moved him aside, took a mop and broom and cleaned up the glass. He grabbed together a ream of wet notes, and announced that the experiment was over.

Father persisted throughout the course to do his own demonstrations and occasionally they kind of worked.

## The Cowboy Experiment

Father's nickname was Joe. Niels Bohr and my father were addicted to western movies while they were in Copenhagen together. All the cowboys in these movies—Gary Cooper types—were called Joe. That's how Father's nickname came about—he was named after a typical cowboy movie hero.

Bohr had some difficulty with cowboy movies. Being a great physicist he took things very literally. After seeing one of the many films in which there was a shootout between a good guy in a white hat and a bad guy in a black hat, Bohr asked Father, "How is it possible that the man in the black hat always reaches for his gun first, but the man in the white hat pulls his gun out even faster and shoots the gun out of his hand?" To you and me the answer is clear: It's the movies! But Bohr said, "I wonder if you are faster when you respond to an incident like a flashing light or somebody moving their hand for their gun than you are when you have to decide yourself when to make a move?" In other words: Is one faster when responding reflexively?

Bohr was quite serious. So, Father and Bohr bought some cap guns and a pair of cowboy hats and set up an experiment. Who would get out the gun and shoot first? Was it the one who drew first or the one who drew in response to the drawn gun? They did the experiment, and as Father said, Bohr in the white hat—the second one to draw—"killed us all." Bohr's hypothesis was actually correct!

Many years after hearing the story, I taught a course called "Molecular Basis of Behavior." Some of the material had to do with the neurological basis of behavior in human beings. In one of the textbooks they described an experiment very similar to Bohr's experiment. Instead of using guns the experimenters used karate fighters. They found that the person who throws the first blow actually has a disadvantage. He or she is slower. The person at which the blow is directed will respond reflexively, and will be much faster.

What can I say? Bohr's and my father's cowboy experiment was ahead of its time. Maybe in a way it was big science.

R. Igor Gamow

# Editor's Note

George Gamow was not only a brilliant physicist, he was a pioneer in the art of science popularization. *Mr. Tompkins in Wonderland* and *Mr. Tompkins Explores the Atom*, originally published in the 1940s, are still classics of the genre today. In those books, Gamow imagined alternate worlds in which the speed of light and Planck's constant were very different from their values in our world and set Mr. Tompkins on an adventure into these strange universes.

In the writings collected in the present volume, Gamow omits the fairy-tale aspect and presents the true laws of physics as we know them in our universe, along with the means of discovery of those laws and their practical consequences for modern technology. Gamow never talks down to the reader, nor does he sidestep difficult issues. Rather, his lucid writing, careful explanations, and vivid imagery make the most abstruse topics accessible.

The first part of this book is reprinted from *The Atom and Its Nucleus*, published in 1961, which sets forth the discovery of the atom and of its inner structure, together with the laws of quantum mechanics that were developed to explain that structure. I have added endnotes and an appendix to bring this section up to date. The second part of the book includes selections from the second edition of *Matter, Earth, and Sky*, published in 1965 (first edition published in 1958). Here

Gamow tackles the physics of our everyday world. These laws are still taught in introductory physics classes much as they were in Gamow's time. This section needed little updating; instead, the endnotes are more in the nature of tangential explorations and expansions of Gamow's discussions on topics of current or historical interest.

Too often we hear laments about the state of science education and awareness in our country. Gamow's exciting and insightful writing is here made available once again as an antidote.

Robert Oerter

# Introduction

## The Development of the Physical Sciences

Since time immemorial man has looked with both admiration and fear at the world into which he was born. He worshipped the Sun and considered it a great benevolent god, giving him light and warmth. (According to a well-known legend, a Greek hero named Prometheus flew up into the heavens and brought back with him some of the divine fire of the Sun to keep his dwellings warm during the cold winter nights, and to make food more tasty and easier to chew by roasting or boiling.) On the other hand, there were hostile gods who liked to scare poor earthlings with loud claps of thunder and terrifying flashes of lightning, or by sending pestilences.

But, among the people for whom various manifestations of nature were nothing but benevolent or hostile acts of gods, there were a few who tried to rationalize all these phenomena and to find intrinsic reasons for them. In olden times they were called *philosophers* or, from the Greek, "lovers of wisdom." One of the greatest philosophers of the ancient world was Aristotle, who lived in the third century B.C. He maintained a large school in Athens, and wrote numerous treatises on logic, psychology, and biology, and also on inorganic phenomena, which he called φυσις (meaning "nature" in Greek). Although the term "physics" is derived from this Greek word, Aristotle himself contributed little to its development, and the early progress of that science was mostly due to other geniuses of the ancient world, such as Pythagoras, who formulated the laws of vibrating strings; Democritus, who conceived the idea of atoms; and Archimedes, the creator of basic laws of mechanics. But the real outburst of physics, and physical sciences in general, started only during the Renaissance and is related to a series of glorious names from Galileo and Newton to Einstein and the other great scientists of the twentieth century. Today physics and the physical sciences in general, such as physical chemistry, astrophysics, geophysics, biophysics, etc., have grown to be a tremendous accumulation of knowledge, exerting a profound influence on the intellectual and technological development of man.

## Our Place in the Universe

In our everyday life we encounter objects of widely differing sizes. Some of them are as large as a barn and others are as small as a pinhead. When we go beyond these limits, either in the direction of much larger objects or in the direction of much smaller ones, it becomes increasingly difficult to grasp their actual sizes. We know that mountains are very large, but at a distance they look quite small, while at short range we can see but a few rocks and cliffs. And bacteria, invisible to the unaided eye, look huge on the microscope.

Objects that are much larger than mountains, such as our Earth itself, the Moon, the Sun, the stars, and stellar systems, constitute what is known as the *macrocosm* (i.e., "large world" in Greek). Very small objects, such as bacteria, atoms, and electrons, belong to the *microcosm* (i.e., "small world" in Greek). If we use the standard scientific unit, a *centimeter* (abbreviated cm; equals 0.3937 inch), for measuring sizes, objects belonging to the macrocosm will be described by very large numbers, and those forming the microcosm by very small ones. Thus, the diameter of the Sun is 139,000,000,000 cm, while the diameter of a hydrogen atom is only 0.0000000106 cm.

Scientist customarily express such numbers in terms of positive or negative powers of 10, and write $1.39 \times 10^{11}$ cm for the diameter of the Sun and $1.06 \times 10^{-8}$ cm for the diameter of a hydrogen atom. Sometimes special very large or very small units are used. Thus, in the macrocosm we use the so-called *astronomical unit* (symbol: AU), which is defined as the mean distance of the Earth from the Sun and is equal to $1.4964 \times 10^{13}$ cm, or a still larger unit known as a *light-year* (symbol: l.y.), which is equal to $9.463 \times 10^{17}$ cm. In the microcosm, we often use *microns* (symbol: $\mu$), defined as $10^{-4}$ cm, and *angstroms* (symbol: $\mathring{A}$), defined as $10^{-8}$ cm.

In Figure I-1, the relative sizes of various objects in everyday life, in the macrocosm, and in the microcosm are shown in a decimal logarithmic scale; i.e., in the scale in which each factor of 10 is represented by one division of the yardstick. The sizes range from the diameter of an electron—and other elementary particles that are about one hundred-thousandth of an angstrom—to the diameter

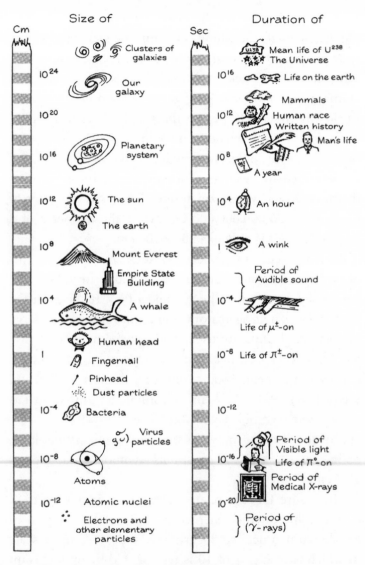

*Figure I-1. Space and time scales of the universe.*

of giant stellar galaxies, which often measure 100,000 light-years across. It is interesting to notice that the size of the human head is just about halfway between the size of an atom and the size of the Sun, or halfway between the size of an atomic nucleus and the diameter of the planetary system (on the logarithmic scale, of course).

Similar vast variations will be found in the time intervals encountered in the study of the microcosm and the macrocosm. In human history, we ordinarily speak about centuries; in geology, the eras are usually measured in hundreds of millions of years, while the age of the universe itself is believed to be over 5 billion years. The revolution period of an electron in the hydrogen atom, on the other hand, is $10^{-15}$ sec (seconds), and the oscillations of particles constituting atomic nuclei have a period of only $10^{-22}$ sec. A comparison (on the logarithmic scale again) of various durations encountered in the macrocosm, microcosm, and in our everyday life is also given in Figure I-1. Notice that the wink of an eye is just halfway between the age of our stellar system and the rotation period of the electron is an atom. Thus, it seems that we are located pretty well in the middle (logarithmically again) between the macro- and microcosm and can look up at stars and down at atoms with equal infer- and superiority.

## Basic Units Used in Physical Sciences

Until the beginning of the nineteenth century, the situation in the field of weights and measures was not much better than the linguistic situation at the Tower of Babel. The units of length varied from country to country, from town to town, from one professor (such as tailor) to another (such as carpenter), and were mostly defined, rather loosely, by reference to various parts of the human body. Thus, an "inch" was defined as a thumb-width, a "hand" or "palm" (still used for measuring the height of race horses) as the breadth of a hand, a "foot" as the length of a British king's foot, a "cubit" as the distance from the elbow to the tip of the middle finger, a "fathom" (used in measuring ocean depths) as the distance between the tips of the middle fingers of the two hands when the arms are outstretched in a straight line, etc. In the year 1791, the French Academy of Sciences recommended the adoption of an international standard of length be based on the size of the Earth. *This unit, called a meter, was to be equal to one ten-millionth of the distance from the pole to the equator.* To prepare a standard meter it became necessary to measure, with all possible precision, at least a part of the earth's meridian, and two French scientists, M. Delambre and M. Mechain, were charged with the task. It took them seven years to measure, by an improved triangulation method, a stretch of meridian from Barcelona in Spain to Dunkirk in Normandy. On the basis of these measurements the academy prepared a "standard meter"—a platinum-irridium bar with two marks on it that was supposed to represent one ten-millionth part of a quarter of the Earth's meridian. The original meter is kept at the Bureau des Poids et Mesures in Sèvres (not far from Paris), and true copies are universally distributed.

Although neither the United States nor Great Britain has accepted the metric system, both recognize the latest international standard of length for the meter, defined by the National Bureau of Standards: 1,650,763.73 wave lengths of the orange-red light of the gas krypton-86 as emitted by a special lamp.

While it is a fact that in stores and factories in the U.S.A., length is customarily measured in yards (yd), feet (ft), and inches (in.), scientific measurements are always expressed in *kilometers* (km; 1,000 m or 0.62 mi, or miles), *meters* (m), *decimeters* (dm; one-tenth of a meter), *centimeters* (cm; one-hundredth of a meter), *millimeters* (mm; one-thousandth of a meter), etc. A peculiar situation exists in the Los Alamos Scientific Laboratory of the AEC in New Mexico where atomic and hydrogen bombs are developed. The nuclear components of the bombs, which involve pure physics, are described in terms of the metric system, while the overall dimensions and the weight are usually given in inches and pounds. It should be mentioned here that the original intention of the French academicians to have a unit of length with a simple relationship to the size of the Earth was not exactly fulfilled. Subsequent, more exact, measurements have shown that the length of a quarter of the Earth's meridian is actually 10,022,288.3 m. The error does not matter, however, as long as we know the exact amount of discrepancy.

Along with the standard unit of length, the metric system also introduced a new unit for the amount of matter, or mass. Disposing of "short and long tons," "pounds (lb)," "ounces (oz)," "drachms," "grains," etc., it uses a *gram* (gm), defined as *the mass of a cubic centimeter (cc or cm³) of water at the temperature (about 4° C) at which it has the greatest density*. A standard kilogram (kg; equal to 1,000 gm) equivalent to the mass of 1 liter (or to one cubic decimeter, 1 cu dm) of water under the above conditions was made of platinum and irridium alloy; the original is kept together with the original meter in Sèvres, and copies are distributed all over the world. While 1gm is the standard unit used in physical measurements, we also use *milligrams* (mg; one-thousandth of a gram) and *micrograms* (μg; one-millionth of a gram) to express the mass of very small amounts of matter.

The clock symbolizes the third fundamental physical unit: a unit of time. A day is divided into 24 hours (hr), and each hour is subdivided into 60 minutes (min), with each minute further divided into 60 seconds (sec). This system of time measurement is based upon that used in ancient Babylon and Egypt, and even the French Revolution (not to mention the Russian one) was unable to convert it into a decimal system. Since we use a decimal system for length and weights, we should logically divide a day into "decidays" (2.4 hr each), "centidays" (8.4 min each), and "millidays" (59.4 sec each). This would necessitate, however, the introduction of "decadays" (10 days each), "hectodays" (3.3 months each), and "kilodays" (2.6 years, or yr, each), and would lead to chaos in speaking about the phases of the Moon or the seasons of the year. In the scientific measurement of time intervals much shorter than a second, however, the decimal system is used, and we speak about *milliseconds* (m-sec; one-thousandth of a second) and *microseconds* (μ-sec; one-millionth of a second).

Having defined the units for length, mass, and time, we can express through them the units for other physical quantities. Thus, the unit of velocity becomes *a centimeter per second* (cm/sec), the unit of material density, *a gram per cubic centimeter* (gm/cm3), etc. This system of units is known as the "CGS (centimeter-gram-second) system" and is always used in scientific literature. Being a unique system accepted by all scientists in the world, it represents a definite advantage over the Anglo-American system of units where the velocity, for example, may be expressed at will in "feet per second," "miles per hour," or even in "furlongs per weekend." Thus, in this book, the metric system will be used exclusively. Readers who are not familiar with meters and kilograms should remember that one *meter* is about equal to one *yard* (1.093611), while one *kilogram* is about equal to two *pounds* (2.20462).

## Derivative Units and the Theory of Dimensions

Accepting three basic units—those of *length* (or distance), *mass*, and *time*—it becomes possible to express through them *all other units* which are encountered in the study of the most complicated physical phenomena, such as the unit of angular momentum, the units of electric charge and current, the unit of magnetic field, the unit of the amount of heat, etc. One of the simplest of these "derivative" units is that of velocity, which is t*he distance covered by a moving body divided by the time necessary to accomplish that motion.* One writes:

$$|\text{velocity}| = \frac{|\text{length}|}{\text{time}}$$

where the vertical lines indicate that we speak about the three physical quantities which are involved, and not about the numerical relationships among their actual values. Similarly, in the case of density we write:

$$|density| = \frac{|mass|}{|length|^3}$$

The way in which any given physical quantity is expressed through units of length, mass, and time is called its *physical dimension* and is independent of the choice of particular units we accept for measuring length, mass, and time. In everyday life we can also use the idea of dimension; for example:

$$|\$| = |\pounds| = |money|$$

meaning that an amount of money expressed in one currency can be translated into any other currency simply by multiplying it by some definite numerical factor.

Familiar examples of formulae with several dimensional quantities are:

$$|salary| = \frac{|money|}{|time|}$$

no matter whether it is dollars per month, or rupees per hour, and:

$$diet = \frac{|food|}{|time| \cdot |consumer|}$$

no matter whether we measure the (nutritious) value of food in kilocalories, time in days, and take for a consumer one single person (as doctors usually do) or use BTU (British Thermal Units), a year, and a regiment of soldiers.

Speaking about thermal phenomena, we will have to introduce two more physical quantities: the amount of heat (which is equivalent to mechanical energy), and the degrees of temperature. As in the course of this volume we progress in the knowledge of various physical phenomena, we will write down the "dimensional formulae" which express various newly-introduced quantities in terms of length, mass, and time.

Dimensional analysis is very important in checking the correctness of various formulae derived in physics. If one writes the equation:

$$A = B$$

where $A$ and $B$ are composed of various different physical quantities, the dimension of $A$ must be the same as the dimension of $B$. If this is not the case, a mistake must certainly have been made in the derivation. Thus:

$$|\text{distance}| = |\text{velocity}| \cdot |\text{time}|$$

or

$$|\text{daily pay}| \cdot |\text{days worked}| = |\text{money}|$$

are dimensionally correct, whereas

$$\frac{|\text{cost per pound}|}{|\text{income}|} = |\text{diet}|$$

is dimensionally wrong.

But, apart from finding mistakes in final formulae, dimensional analysis has a much more useful application in the process of the derivation of the formulae themselves. We will give several examples of this later in this book.

## *Observation and Experiment Versus Theory*

Although the physical sciences study the phenomena in the world in which we live, most of the time we remain just spectators of what is happening around us. For example, we observe that the Sun moves daily across the sky, and we can measure the amount of light and heat it supplies to the Earth. But we cannot modify these phenomena, and in this sense astronomy remains an observational science. Even the space rockets flying between the planets of the solar system only improve observation; they cannot change the course of things. Geophysics in general, and meteorology in particular, were until quite recently also purely observational sciences. Today, however, we can carry out experiments by seeding the clouds to see if that will cause them to burst into rain, and we can explode nuclear weapons well beyond the terrestrial atmosphere to find out what effect they have on radio transmission, etc. One even hopes that, after acquiring sufficient knowledge about atmospheric phenomena, we will be able to influence them at will, thus disproving Mark Twain's statement that "Everybody talks about the weather, but nobody does anything about it." Physics, which deals with smaller objects than the Sun or the Earth, became an experimental science a long time ago; today we can carry out various controlled experiments, either as simple as measuring the volume of a given amount of water when it is heated, or as complex as shooting artificially accelerated high-energy particles at the nuclei of various atoms.

The observational and experimental physical sciences collect tremendous amounts of empirical material which have to be carefully analyzed, classified, and ordered. As a result of these efforts, scientists establish various *empirical laws*, such as those describing planetary motion (*Kepler's Laws*); the relation between the angles of the incidence and the refraction of light entering from air into glass (*Snell's Law*); proportionality between electric and heat conductivities (*the Wiedemann-Franz Law*); or the dependence of the lifetime of radioactive bodies on the energy of the particles they emit (*the Geiger-Nuttall Law*). Now to the stage come theoreticians who, though being not too experienced in observational and experimental techniques, are nevertheless well trained in mathematics, which is, in this respect, a shorthand (or rather "shortbrain") art of investigating complicated correlations between various quantities and finding the hidden relations between them. Thus, for example, Newton's Theory of Universal Gravity explained in one swing Kepler's Law of Planetary Motion, the precession of equinoxes, the ocean tides, etc. The Electron Theory of Metals explained the relation between electric and heat conductivity, along with many other phenomena, while wave mechanics, which explained many previously mysterious atomic phenomena, led to a simple understanding of the radioactive decay of atomic nuclei.

The history of physics and of the physical sciences in general can be considered as a kind of game between the observationalists and the experimentalists on one side and the theoreticians on the other. Sometimes the former are way ahead; sometimes the latter. For example, toward the end of the last century, a tremendous amount of empirical material concerning atoms had been collected by generations of experimental physicists and chemists, but there was no understanding of the laws concerning atoms. At the turn of the century, the German theoretical physicist, Max Planck, formulated a bold hypothesis: that energy can be available only in certain well-defined portions known as *quanta*. Within only three decades after this breakthrough, a host of brilliant theoretical physicists (whose names will be referred to later in this book) completely disentangled the riddle of the atom and its nucleus, uniting into a harmonious theory all the empirical results obtained before. In the early 1930's, experimentalists made exciting new advances by discovering a large number of previously unknown particles (muons, pions, kaons, hyperons, etc.) while theory came almost to a standstill. And this is the situation today and will be until, due to some unconventional idea proposed by a theoretician— who may now be just a freshman in some university—will start a new avalanche of theoretical developments.

# PART 1

# The Atom

# CHAPTER 1

# The Atom in Philosophy and Chemistry

## *The Greek Idea*

The ancient Greek philosophers who speculated about the nature of things suspected that the *immense variety of different substances forming the world results from a combination of comparatively few simple elements.* Democritus (fifth century B.C.) believed that there are four elementary substances: air, water, stone, and fire, all formed by a very large number of very small particles called atoms, i.e., "indivisibles" in Greek. The atoms of air were supposed to carry the properties of "lightness" and "dryness," the atoms of water the properties of "heaviness" and "wetness," the atoms of stone the properties of "heaviness" and "dryness," while the atoms of fire were supposed to be very mobile, "slippery, and hot." On the basis of these ideas, the Greek philosophers attempted to explain the various transformations of matter as resulting from the reshuffling of the atoms constituting matter. They believed that the material of a growing plant is composed of water and stone atoms provided by the soil and atoms of fire supplied by the rays of the sun. In modern chemical terminology, the Greek formula for wood would be SWF. The drying of wood was considered to be the escape from the wood of water atoms, SWF   SF + W, and the burning of wood the decomposition of dry wood into fire atoms (flame) and stone atoms (ashes), SF   F + S. Metals were considered to be the combination of stone atoms with varying amounts of fire atoms, SFn (the fire atoms were supposedly responsible for metallic glitter). Iron was supposed to be rather poor in fire atoms, but gold was considered to have the maximum amount of them. The formation of metals from ores treated in a furnace was thought to result from the union of the stone atoms of the ore and the fire atoms of the flame, $S + nF$   $SFSFn$, and it seemed logical to expect that by enriching common metals like iron or copper with fire atoms, one should be able to turn them into gold. This point of view, which also prevailed in the Middle Ages, explains the incessant efforts of medieval alchemists to transform common metals into precious ones.

We know now that these views were quite wrong. The metals themselves and not their ores are elementary substances, and the process that takes place in blast furnaces does not add fire atoms to stony ores and turn them into metals but, quite on the contrary, subtracts oxygen from metallic oxides (ores) and thus liberates pure metals. Also, the material of a growing plant is obtained by the carbon dioxide (carbon + oxygen) from the air combining with the water (hydrogen + oxygen) from the soil, while the sun's rays supply only the energy necessary for synthesizing complex organic substances from these simple ingredients. The difference between the ancient and the modern view in chemistry is shown in Figure 1.1. Although the attempted explanations were completely wrong, the idea of reducing the multitude of chemical substances to combinations of comparatively small numbers of simple elements was basically correct and now lies at the foundation of modern chemistry.

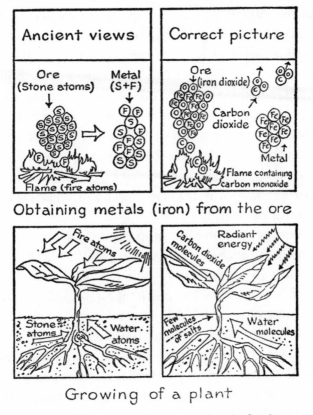

Figure 1.1. *Two wrong vs. two correct views in chemistry.*

## Atomic Weights and Valency

The most important fact concerning the formation of various chemical compounds from elements is contained in the so-called law of constant proportions, which states that *the relative amounts of different chemical elements needed to form a definite chemical compound always stand in a certain given ratio.* Thus, when we place a mixture of hydrogen and oxygen gases in a thick-walled container and ignite the gases with an electric spark, we produce a rather violent chemical reaction (or explosion) which results in the formation of water. If the original proportions of hydrogen and oxygen are 1:8 by weight, the reaction will be complete and there will be nothing left over of either of the two gases. If, however, there is originally more hydrogen or more oxygen than is specified in the 1:8 proportion, then a corresponding excess of either gas will be left over. (There exists, however, another compound of hydrogen and oxygen known as *hydrogen peroxide* in which the ratio of the two elements is 1:16.)

The law of constant proportions was interpreted by the British chemist, John Dalton (1766-1844), as being due to atom-to-atom union in the formation of chemical compounds. To explain the above-described facts concerning water and hydrogen peroxide, one can assume that the weight ratio of the atoms of hydrogen and oxygen is 1:16 and that there is one atom of hydrogen per each atom of oxygen in hydrogen peroxide, while there are two hydrogen atoms per each oxygen atom in the case of water. Therefore, writing H for a hydrogen atom and O for an oxygen atom and using a subindex to denote the number of atoms of each, we can express the chemical composition of these two substances as:

water molecule = $H_2O$
hydrogen peroxide molecule = HO (or $H_2O_2$, as it can be shown to be by other methods)

The second way of writing the expression for hydrogen peroxide indicates that this molecule has one oxygen atom too many in comparison with the much more common compound, water. And, indeed, hydrogen peroxide is an unstable substance that decomposes spontaneously according to the equation:

$$H_2O_2 \rightarrow H_2O + O$$

The free oxygen atoms that are liberated in this reaction possess strong oxidative properties, which make $H_2O_2$ useful in various bleaching processes, not the least of which is the turning of a dark-haired girl into a platinum blonde.

Similarly, the union of carbon and oxygen may result either in carbon dioxide, $CO_2$, or, in the case of burning with an unsufficient supply of oxygen, in carbon monoxide, CO. In contrast to hydrogen peroxide, CO molecules lack one oxygen atom and are anxious to rob that extra oxygen atom from any other molecule which does not hold it strongly enough. The ratio by weight of carbon to oxygen is carbon monoxide is 3:4, which can also be written as 12:16. Since the atomic weight of oxygen was established as 16 (i.e., it weighs 16 times as much as a hydrogen atom, which for the present we can consider to be of unit weight), the

atomic weight of carbon must be 12. Carbon also unites
with hydrogen, giving rise to a gas known as *methane* or
"marsh gas." The ratio of hydrogen to carbon in methane
is 1:3 or 4:12, and, since 12 is the weight of one carbon
atom, the formula of methane must be $CH_4$. Let us now
consider a slightly more complicated example presented by
an analysis of ethyl alcohol, which is 52.2 percent carbon,
34.8 percent oxygen, and 13.0 percent hydrogen. By
noticing that the ratio $^{52.2}/_{34.8}$ is 1.50, whereas the ratio of
the atomic weights of carbon and oxygen is only 0.75, we
can conclude that there must be two carbon atoms for each
oxygen atom. If there were only one hydrogen atom for
each oxygen atom, the ratio of corresponding percentages
would have to be $^1/_{16} = 0.0625$, but the ratio is actually
$^{13.0}/_{34.8} = 0.375$, i.e., six times larger. Therefore there must
be six hydrogen atoms per oxygen atom, and the formula
for ethyl alcohol is $C_2OH_6$.

The ability of atoms to unite with one or more other
atoms is known as *chemical valency* and can be represented
in an elementary way by drawing on each atom a number
of hooks that can be coupled with the hooks of other
atoms. In the examples so far considered, we have ascribed
to hydrogen atoms a valency of 1, to oxygen 2, and to
carbon 4. The way atoms are then bound into molecules
(the so-called structural formula of the molecule) is shown
in Table 1-1.

Valence "hooks" can also act between identical atoms
and bind them into "diatomic" or "triatomic" molecules of
a simple chemical substance, as indicated in the last three
items of Table 1-1. Similar relations can be found for other
chemical elements and for more complicated chemical
compounds.

TABLE 1-1  MOLECULAR STRUCTURE OF VARIOUS COMPOUNDS

| Water | | H—O—H |
|---|---|---|
| Hydrogen Peroxide | | H—O—O—H |
| Carbon Dioxide | | O=C=O |
| Carbon Monoxide | | =C=O |
| Methane | | H<br>\|<br>H—C—H<br>\|<br>H |
| Ethyl Alcohol | | H  H<br>\|  \|<br>H—C—C—O—H<br>\|  \|<br>H  H |
| Hydrogen Gas | | H—H |
| Oxygen Gas | | O—O |
| Ozone | | O<br>/ \\<br>O—O |

In speaking about chemical valency, we must mention six very peculiar elements: argon, helium, krypton, neon, radon, and xenon. These do not possess any chemical valency whatsoever. The atoms of these elements despise any chemical intimacy and prefer to remain alone; they do not even form pairs between themselves as other atoms often do, so their molecules are always "monatomic." Closely connected with this chemical inertness is the fact that all these six substances are gases and liquefy only at very low temperatures. Using the self-apparent analogy, we call these elements *noble gases* or, sometimes, *rare gases*, since, indeed, they all are rather rare on the earth. As everybody knows, helium is used for filling balloons and dirigibles to avoid fires, and neon, which emits a brilliant red light when subjected to an electric discharge, is used for making luminous signs for advertising.

## The Periodic Law

Although the arrangement of chemical elements in alphabetical order is convenient for inventory purposes, it is more reasonable to arrange them in the order of increasing atomic weights. In doing so, we find rather remarkable regularities which have led chemists to a rational classification of the elements. Arranging the elements in order of atomic weights,* we obtain the following *sequence*: H, *He*, Li, Be, C, N, O, F, *Ne*, Na, Mg, Al, Si, P, S, Cl, *A*, K, Ca, Sc, Ti, V, Cr, Mn, Fe, Co, Ni, Cu, Zn, Ga, Ge, As, Se, Br, *Kr*, Rb, Sr, Y, Zr, Nb, Mo, Tc, Ru, Rh, Pd, Ag, Cd, In, Sn, Sb, Te, I, *Xe*, Cs, Ba, La, etc. We notice, first of all, that there is a remarkable regularity in the distribution of noble gases, shown in *italics*, throughout the sequence: there is only 1 element preceding *He*, 7 elements between *He* and *Ne*, another 7 elements between *Ne* and *A*, 17 elements between *A* and *Kr*, and another 17 elements between *Kr* and *Xe*. Finally, there are 31 elements between *Xe* and *Rn*, which is the heaviest known noble gas.

---

*The careful student will notice that K, Ni, and I are out of order, but, as it was found later, the sequence of chemical properties has priority over atomic weights.

The elements immediately following the noble gases, lithium, sodium, potassium, rubidium, and cesium, are physically and chemically very similar to each other. They are all light, silvery-white metals with high chemical activity. If we drop a small piece of any of these elements in water, it will undergo a violent chemical reaction of the type:

$$Li + H_2O \rightarrow LiOH + H$$
$$Na + H_2O \rightarrow NaOH + H$$
*etc.*

liberating hydrogen and forming the corresponding "hydroxide" with water (structural formula, Li—O—H, etc.). The hydrogen liberated in this reaction often becomes ignited and produces a flame which takes on the characteristics color of the vaporized metal (yellow for sodium, red for potassium, etc.). Uniting with hydrogen and oxygen, these elements form "hydrates" and "oxides" of the type LiH (Li—H), $Li_2O$ (Li—O—Li), etc., showing that their valency is 1. These elements are commonly known in chemistry as *alkali metals*.

The second neighbors to the right of the noble gases, beryllium, magnesium, calcium, strontium, barium, and radium, also form a homologous group known as alkali-earth metals. As their name indicates, they are similar to the alkali metals, but, as a rule, they are much harder and less reactive. Reacting with water, they produce compounds of the type $Ca(OH)_2$ (H—O—Ca—O—H), while uniting with hydrogen and oxygen they give rise to compounds such as $CaH_2$(H—Ca—H) and CaO (Ca=O), which indicates that their valency is 2. Similarly, we find that the third group to the right, boron, aluminum, etc., possesses a valency of 3 as demonstrated by such compounds as boron oxide, $B_2O_3$(O=B—O—B=O), and aluminum hydroxide, $Al(OH)_3$.

Now if we look at the elements standing to the left of the noble gases, we will find that they are very similar to each other, but as different from metals as they could possibly be. This group comprises fluorine, chlorine, bromine, iodine, and astatine, and they are known as the halogens. They have a strong affinity for both alkali and alkali-earth metals, with which they form such compounds as NaCl (ordinary table salt) and $CaBr_2$, indicating that they possess a single valency. The second neighbors to the left of the noble gases, oxygen, sulfur, etc., are also in some ways similar to each other and possess a valency of 2.

The existence of homologous groups and of a certain periodicity in the chemical properties of elements arranged in the order of increasing atom weights was noticed by several chemists during the nineteenth century, but the most important step of actually arranging the elements into a periodic table was made in 1869 by the Russian chemist, Dmitri Mendeleev (1834-1907). Mendeleev was handicapped in his studies because in his time the list of known chemical elements was rather incomplete and, in particular, the existence of the noble gases was not even suspected. From the sequence given above, Sc, Ga, Ge, Tc, and Rh were missing, making the sequence quite irregular except for the first two periods. Driven by a deep belief that there *must be* a regular periodicity in the natural sequence of elements, Mendeleev made the bold hypothesis that the deviations from the expected periodicity in his list were due to the failure of contemporary chemistry to have discovered some of the elements existing in nature. Thus, in constructing his table, he left a number of empty spaces to be filled in later by future discoveries. He gave to the "missing elements" names formed by adding the prefixes *eka* or *dvi*, meaning "first" and "second" in Sanskrit, to the names of neighboring homologous elements. In certain instances, he also reversed the atomic-weight order of elements in order to comply with the demands of the regular periodicity of their chemical properties. Using his table, shaky as it was, he was able to predict the physical and chemical properties of six "missing elements" on the basis of the known properties of their alleged neighbors. He called these elements eka-boron, eka-aluminum, eka-silicon, eka-manganese, dvi-manganese, and eka-tantalum. His predictions turned out to be in excellent agreement with the actually observed properties of the "missing elements" when they were finally found and named: scandium, gallium, germanium, technetium,* rhenium, and polonium. Just as an example, we give in Table 1-2 the comparison of Mendeleev's predictions of the properties of his hypothetical element "eka-silicon," with the actually observed properties of this element, which was found fifteen years later by a German chemist, Winkler, and given the name germanium.

Pretty good for a prediction at this stage in the development of chemistry!

By enumerating the elements from 1 (for hydrogen) and up as they come in the periodic system of elements, we obtain what is known as the *atomic numbers* of the elements.

---

*Technetium, an unstable element normally non-existent in nature, was produced only recently in atomic piles.

**TABLE 1-2**

| Mendeleev's prediction for eka-silicon (Es) (1871) | Winkler's data for germanium (Ge) (Discovered in 1886) |
|---|---|
| Atomic weight will be about 72 | Atomic weight is 72.6 |
| Will be obtained from $EsO_2$ or $K_2EsF_6$ by reduction with Na | Was obtained from $K_2GeF_6$ by reduction with Na |
| Will be a dark gray metal with high melting point and density about 5.5 | Is a gray metal with melting point 958° C and density 5.36 |
| On heating, Es will form the oxide $EsO_2$ with high melting point and density 4.7 | Reacts with oxygen forming $GeO_2$ with melting point 1,100°C and density 4.7 |
| The sulfide $EsS_2$ will be insoluble in water but soluble in ammonium sulfide | $GeS_2$ is insoluble in water but readily soluble in ammonium sulfide |

Thus, the atomic number of carbon is 6, that of mercury is 80, and that of mendelevium, 101. The atomic numbers of the six noble gases that form important landmarks of chemical periodicity are: 2, 10, 18, 36, 54, and 86. It is convenient to represent the periodic system of elements by a three-dimensional spiral structure that is shown in Figure 1-2. The backbone of the structure is the column containing the noble gases running all the way from He down to Rn. The next column to the right contains the alkali metals, with hydrogen placed at the top because its chemical properties are similar to those of the alkali metals. To the left and around the corner from the noble gas column is the one containing the halogens.

The first two periods, from He to F and from Ne to Cl, contain 8 elements each and fall neatly into this scheme, but the next period contains 18 elements and constitutes a problem. On the basis of chemical properties, there seems to be no doubt that the 3 elements that follow A (K, Ca, and Sc) must be placed under the 3 corresponding elements (Na, Mg, and Al) of the previous period and that those preceding Kr (As, Se, and Br) should be under those preceding A (P, S, and Cl), but we do not seem to have places for the remaining 11 elements Ti to Ge). To dispose of this difficulty, we place Ti and Ge, which both resemble Si, under that element and make an extra loop to accommodate the remaining 9 elements (V to Ga). The same situation arises in the next and in all of the following periods so that the extra loop perpetuates itself all the way to the end of the known sequence of elements. In the beginning of the fifth period, we encounter further trouble of the same kind and are forced to build another extra loop to accommodate 14 extra elements (Ce to Lu), known as the *rare earths*. The sixth and last period runs in the same way with most of the natural and artificial radioactive elements forming a loop under that formed by the rare earths.

Things become quite complicated, and Dmitri Ivanovich Mendeleev would probably be horrified by the looks of it, but that's how it is. Nevertheless, in spite of the complexity of the diagram (which reflects the complexity of the internal structure of the atom), the periodic system of elements in Figure 1-2 gives a very good representation of the properties of the different elements.

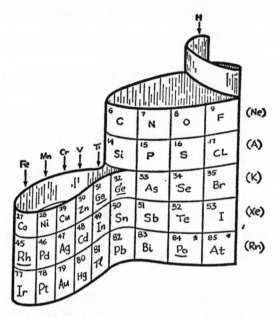

*Figure 1-2. The periodic system of the elements represented as a wound ribbon. The diagram on the next page shows the other side of the second loop. At present the ribbon is cut at atomic number 101 (mendelevium). An asterisk indicates that the element is unstable (radioactive), and an asterisk in parenthesis indicates the presence of a radioactive isotope in the normally stable element. The properties of the underlined elements were predicted by Mendeleev.*

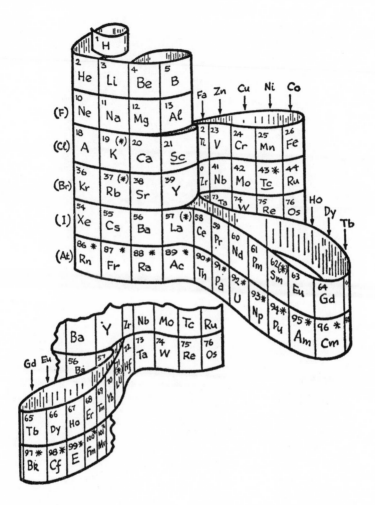

# CHAPTER 2

# The Electric Nature
# of Matter

## *Positive and Negative Ions*

Pure distilled water is a very poor conductor of electricity. However, if we dissolve in water a small amount of some acid or salt, its electrical conductivity becomes quite appreciable. In contrast to the case of metallic conductors, the passage of electric current through water solutions is associated with certain chemical phenomena, the nature of which depends on the particular solute used. If we pass an electric current through a solution of nitric acid ($HNO_3$), small gas bubbles will be formed on both electrodes and will gradually rise to the surface. We can collect these gases in two long inverted glass cylinders that are placed about the electrodes and that are originally completely filled with water (Figure 2-1 $a$). When we analyze the gas liberated on the negative electrode, we find it to be hydrogen; in fact, if we open the valve at the top of the glass cylinder placed above this electrode, we can ignite the gas streaming out from it, and in the process of burning, the hydrogen will unite with atmospheric oxygen and form water vapor. The gas that is collected in the cylinder placed above the positive electrode is oxygen; if we open the valve at the top of that cylinder and place a burning match into the stream of outcoming gas, it will flare up more intensely because of the additional oxygen supply.

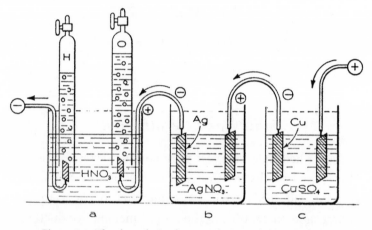

*Figure 2-1. The electrolysis of water solutions of nitric acid (a), silver nitrate (b), and copper sulfate (c) by the same current.*

Thus, the passage of electric current decomposes water into its two elementary constituents, hydrogen and oxygen. How does this happen, and why should it require something dissolved in the water to get things going? The water molecule, $H_2O$, is quite tightly bound together and has very little tendency to break apart into ions. In fact, at room temperature only about 1 molecule in $10^7$ will be split up into $H^+$ and $OH^-$ ions. Such a small number of ions migrating through an electric field constitutes so small a current that pure distilled water may be considered a fairly good insulator. However, the nitric acid molecules (as well as the molecules of salts and bases) split up readily into $H^+$ and $NO_3^-$ when dissolved in water. Thus a large number of charged ions are provided, and their migration in the electric field between anode and cathode can constitute a large current.

When an electric potential is applied to the cathode and anode, the positive ions of hydrogen are attracted to the negative electrode (cathode) and the negative $NO_3^-$ ions are attracted to the positive electrode (anode). As the result of the ionic motion caused by these attractions, an electric current flows through the water solution, a current that would have been impossible in pure water. When the traveling ions $H^+$ and $NO_3^-$ arrive at their respective electrodes, they release their electric charges into the metal; hydrogen rises to the surface in the form of tiny bubbles, while the neutral $NO_3$ reacts with water according to the equation $2NO_3 + H_2O \rightarrow 2HNO_3 + O$, liberating free oxygen and regenerating the original molecules of nitric acid. (There are secondary reactions producing oxides of nitrogen, but we may leave these complications to the chemists.) Thus, the passage of electric current through the water resulted in nothing more than the breaking up of the water molecules into their hydrogen and oxygen components.

If, instead of using nitric acid, we use one of its salts (in which hydrogen is replaced by a metal), the metal will be deposited on the surface of the negative electrode. When, for example, we pass an electric current through a solution of silver nitrate, $AgNO_3$ (Figure 2-1*b*) we will notice that after a while the cathode will be covered with a thin layer of silver. This method of coating surfaces with thin layers of various metals is known as electroplating and has many useful and practical applications. Just as in the case of nitric acid, the electrolytic process in the silver nitrate solution is due to the fact that the molecules of this salt break up into two oppositely charged ions, $Ag^+$ and $NO_3^-$, which are driven in opposite directions by the applied electric potential. When the $Ag^+$ ions reach the negative cathode, they pick up their lost electrons from the cathode and become neutral insoluble Ag atoms which form the cathode deposit. At the anode, the $NO_3^-$ ions give up their electrons to the anode, so the net effect is that of a stream of electrons flowing within the electrolysis tank from cathode to anode.

## The Laws of Faraday

Michael Faraday, whose name is associated with the theory of electric and magnetic fields, was the first to investigate in detail the laws of electrolytic processes. He found first of all that, for each given salt solution, the amount of material deposited at the electrodes is directly proportional to the strength of the electric current and to its duration, or, in other words, that *the amount of material deposited on the electrodes is directly proportional to the total amount of electric charge which had passed through the solution.* From this first law of Faraday, we conclude that each ion of a given chemical substance carries a well-defined electric charge.

In further studies, Faraday investigated the relative amounts of electric charge carried by ions of different chemical substances. To compare these amounts, he passed an electric current consecutively through the solutions of several different substances, such as nitric acid, silver nitrate, and copper sulfate, as is shown in Figure 2-1a, b, and c. In the case of nitric acid, a certain amount of hydrogen gas was liberated on the cathode, while a certain amount of silver was deposited on the cathode in the case of the silver nitrate solution. Faraday measured the amounts of hydrogen and silver produced in these experiments and found that the ratio of the weight of deposited silver to the weight of liberated hydrogen was 107.02. Chemists had before this time determined from many ingenious experimental measurements the relative weights of the atoms of the chemical elements. These relative *atomic weights* were arranged in a table in which the weight of the oxygen atoms in the atmosphere was arbitrarily taken as 16.000, and the weights of all the other kinds of atom were expressed in units of $1/16$ of the weight of the oxygen atom; 107.02 is exactly the ratio of the atomic weight of silver to the atomic weight of hydrogen. Thus, Faraday concluded that the same number of atoms of Ag and H had been deposited and that *one ion of silver carries exactly the same electric charge as one ion of hydrogen.* It would be premature, however, to conclude that all ions carry the same electric charge. In fact, comparing the amount of silver liberated in the electrolysis of silver nitrate with that of copper liberated by the same electric current flowing for the same length of time in the electrolysis of copper sulfate, we find that the weight ratio of silver to copper is 3.40 instead of the 1.70 ($^{107.9}/_{63.5}$) that would correspond to one atom of silver per atom of copper. Notice, however, that 1.70 is exactly one-half of 3.40, and if we write the observed ratio in the form ($2 \times ^{107.9}/_{63.5}$), we conclude that *one ion of copper carries twice as much electricity as one ion of silver.* We can interpret this by saying that the silver ion has lost one electron, while the copper ion has lost two electrons and therefore has a double positive charge. The number of electrons lost or gained by an ion is one aspect of what the chemist calls *valence*. Thus, hydrogen, silver, and the nitrate group ($NO_3$) have a valence of 1 (monovalent), copper and the sulfate group ($SO_4$) have a valence of 2 (divalent), whereas aluminum ions have a valence of 3 (trivalent).

Thus with several electrolytic cells in series, as in Figure 2-1, for each atom of a monovalent element that is deposited, only $1/2$ of a divalent atom, or $1/3$ of a trivalent atom can be deposited. Chemists call the atomic weight divided by the valence the *equivalent weight*, and Faraday's second law of electrolysis states that *when the same amount of electric charge flows through different electrolytic cells, the amounts of the substances deposited (or liberated) are in direct proportion to their equivalent weights.*

For example, we can place two cells in series (which guarantees that the same amount of charge will flow through each), one cell containing silver nitrate and the other gold chloride (gold is trivalent), and allow current to flow until we have 1.00 gm of silver deposited on the cathode of the first cell. At this time, how much gold will have been deposited on the cathode of the other cell? The equivalent weight of silver, since silver in monovalent, is the same as its atomic weight, or 107.9. Gold has an equivalent weight of $197.0/3 = 65.7$. Therefore, we can write:

$$\frac{\text{wt Ag deposited}}{\text{wt Au deposited}} = \frac{1.00}{x} = \frac{107.9}{65.7}$$

or:
$$x = 0.609 \text{ gm Au deposited}$$

It has been found that the passage of 96,500 coulombs of charge will deposit a mass, in grams, of any element which is numerically equal to its equivalent weight. (This amount of any element is more formally called a *gram-equivalent weight*; one *gram-atomic weight* is, of course, an amount of substance whose mass in grams equals its atomic weight.)

## The Passage of Electricity through Gases

The next step in the study of the electric nature of matter was made by J. J. Thomson (1856-1940), another famous Britisher. While Michael Faraday studied the passage of electric current through liquids, J. J. (as he was known to his colleagues and his students) later concentrated his attention on the electrical conductivity of gases.

When we walk in the evening along the downtown streets of a modern city, we observe the bright display of neon (bright red) and helium (pale green) advertising signs. Modern offices and homes are illuminated by fluorescent light tubes. In all these cases, we deal with the passage of high-voltage electric current through a rarefied gas—the phenomenon that was the object of the lifelong studies of J. J. Thomson. As in the case of liquids, the current passing through a gas is due to the motion of positive and negative ions driven in opposite directions by an applied electric field. The positive gas ions are similar to those encountered in the electrolysis of liquids (being the positively charged atoms or molecules of the substance in question), and the negative ions in this case are the much less massive singly charged particles that we now know to be electrons.

To study these, at that time, mysterious particles Thomson, in 1897, used an instrument shown schematically in Figure 2-2. It consisted of a glass tube containing highly rarified gas with a cathode placed at one end of it and an

anode located in an extension on the side. Because of this
arrangement, the negative ions, which form the "cathode-
rays" that move from left on right in the drawing, miss
the anode and fly into the right side of the tube. The
tube broadens here, and its flat rear end is covered with
a layer of fluorescent material which becomes luminous
when bombarded by fast-moving particles. This tube is
very similar to a modern TV tube where the image of
pirouetting ballerinas or sweating prize fighters is also
due to the fluorescence produced by a scanning electron
beam. But in those pioneering days of what we now call
*electronics*, one was satisfied with much simpler shows;
placing a metal cross in the way of the beam, Thomson
observed that it cast a shadow on the fluorescent screen,
indicating that the particles in question were moving along
straight lines, similar to light rays.

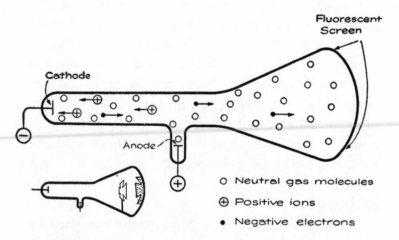

*Figure 2-2. The passage of electric current through rarefied
gas.*

## *The Charge-to-Mass Ratio of an Electron*

Thomson's next task was to study the deflection of the beam caused by electric and magnetic fields applied along its path. Indeed, since the beam was formed by a swarm of negatively charged particles, it should be deflected toward the positive pole of the condenser that produces the electric field shown in Figure 2-3a. On the other hand, since a beam of charged particles is equivalent to an electric current it should be deflected by a magnetic field directed perpendicularly to its track (Figure 2-3b) according to the laws of electromagnetic interactions.

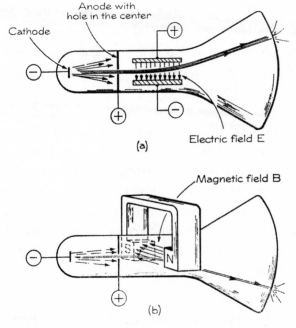

*Figure 2-3. J. J. Thomson's method of measuring the velocity and the mass of electrons: (a) the electric deflection of an electron beam and (b) the magnetic deflection of such a beam.*

The deflection of a particle will depend, of course, on how much force is applied to it. For a charged particle in an electric field, the force depends only on the particle's charge and on the strength of the field. For a magnetic field, however, the situation is different; a magnetic field has no effect on a stationary charge but it does exert a force on an electric current, which is nothing more than a stream of *moving* charges. Hence the deflection in this case will depend on the strength of the magnetic field, the charge on the moving particle, and also on its velocity.

By combining the two experiments shown in Figure 2-3*a* and *b*, Thomson was able to get valuable information about the little negatively charged particles called *electrons*. In a tube equipped with *both* an electric and magnetic field, Thomson adjusted the strengths of the two fields so that the beam of electrons continued straight ahead without any deviation.

It can be shown from the theory of electric and magnetic deflections of a moving charge that the ratio $e/m$ of the electron's charge to its mass can be obtained from the measured strengths of the fields. Thomson's experiments, and those of later workers, give the value $e/m = 1.76 \times 10^{-8}$ coulombs/gram. Unfortunately, however, he was not able to solve his equations to determine their electric charges, because the deflections of the electron beams depend also on the mass of the electrons, which he did not know.

## *The Charge and Mass of an Electron*

This work of Thomson's paved the way for the work of the celebrated American physicist, Robert A. Millikan, who directly measured the charge of the electron by means of a very ingenious experiment illustrated in Figure 2-4. A cloud of tiny oil droplets was sprayed into the space above the plates, and a small hole in the top plate was uncovered long enough for one of the droplets to drift down through the hole into the space between the plates, where it could be observed through a microscope set into the wall of the vessel. By means of a relationship known as *Stokes' law*, the weight of a small droplet can be determined from the rate at which it settles downward through the air. Millikan could measure the rate of settling with no electric field between the plates and thus compute the weight of the droplet.

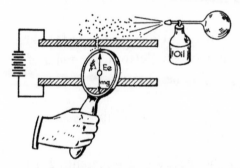

*Figure 2-4. A schematic arrangement of Millikan's experiment for measuring elementary charge.*

Ultraviolet light can pull electrons away from the molecules of objects on which it falls, so by allowing a beam of ultraviolet light to shine between the plates, Millikan could cause the droplet to have a slight charge that could change suddenly from time to time as it collided with charged air molecules. By varying the potential applied across his plates, he could adjust the electric field until the droplet would hang motionless, neither rising nor falling. Under these equilibrium conditions, the upward force caused by the electric field was just equal to the weight of the droplet, and thus:

$$Eq=mg$$

from which $q$, the charge on the droplet, could be easily figured.

It turned out that all the charges measured in this way were small integral multiples of a certain quantity that was apparently the elementary electric charge, or the charge of an electron. Numerically he found that the value of this elementary charge is $1.60 \times 10^{-19}$ coulomb, or $4.80 \times 10^{-10}$ esu.

From Thomson's charge-to-mass ratio and a direct knowledge of the charge on an electron, the mass of an electron can be computed to be:

$$\frac{1.60 \times 10^{-19} \text{ coulombs}}{1.76 \times 10^{8} \text{ coulombs} / \text{gram}} = 9.11 \times 10^{-28} \text{ grams}$$

The discovery of the electron as representing a free electric charge and the possibility of its extraction from neutral atoms was the first indication that *atoms are not indivisible particles but complex mechanical systems composed of positively and negatively charged parts.* Positive ions were interpreted as having a *deficiency* of one or more electrons, whereas negative ions were considered as atoms having an excess of electrons.

## Canal Rays and Isotopes

While the study of cathode rays in Thomson's tube led to the discovery of electrons, *canal rays*, which are a stream of positively charged gas ions, were also very helpful for the understanding of the inner nature of the atom. The apparatus Thomson used for the study of canal rays was a modification of the tube used for determining the *e/m* ratio, and is shown in Figure 2-5. A small amount of gas is left within the tube, and when a swiftly moving electron collides with a gas molecule, an electron is likely to be knocked off, making the molecule into an ion with a positive charge. The mass and electric charge of these positively charged canal rays can be analyzed by deflecting them in electric and magnetic fields. Thomson used parallel electric and magnetic fields, so that passing ions received thrusts in both vertical and horizontal directions. Although we will not stop to do it, it is not difficult to show that a stream of positive ions, all of the same charge and mass, will leave a trace in the shape of a parabola on the fluorescent end of the tube. Those ions having high speeds will be deviated little; slower ions will strike farther out on the parabola. The mass of the ions can be computed quite accurately from the geometry of the parabola.

In measuring the mass of the particles forming canal rays in a tube filled with neon gas, Thomson expected to confirm the chemical value of the atomic weight of neon, which was known to be 20.183. However, instead of this value he got only 20.0, which was considerably lower and well beyond the limits of possible experimental error. The discrepancy was explained when Thomson noticed that the beam of neon ions passing through the magnetic and electric fields was not deflected as a single beam, but was split into three branches (Figure 2-5). Thomson's co-worker, F. W. Aston, later used more complex, improved equipment, and found a very faint third branch, which is also shown in the drawing. The particles in the main branch, containing over 90.5 percent of all the neon ions, had a mass value of 20.0; the other fainter branch contained 9.2 percent, and had a mass of 22.0, and a still fainter branch containing 0.3 percent of mass 21.

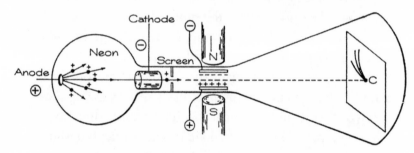

*Figure 2-5. The apparatus that led to the discovery of isotopes. Positive ions of neon were accelerated by an electric field and formed a thin beam after passing through a slit in a screen. The beam was deflected by a combined electric and magnetic field and fell on the fluorescent screen at the far right end of the tube. If all Ne ions had the same mass, though different veloci-ties, the line on the screen would be in the shape of a parabola. But there were three different parabolas corresponding to the masses 20, 21, and 22.*

This was very remarkable! Here Thomson had found two kinds of neon atoms, *identical in chemical nature and in their optical spectra, but different in mass.* On top of this, both mass values were almost exactly integral numbers. Ordinary neon, then, was actually a mixture of two different neons (three, in fact, as was discovered later) and the chemical weight was just the average weight of this mixture.

The different types of neon were called *isotopes* of this element which means in Greek "same place" and refers to the fact that all the neons of different weight occupy the same place in the table of the elements. We usually denote isotopes by placing an index indicating the mass at the upper-right corner of the symbol of the element; thus $Ne^{20}$, $Ne^{21}$, and $Ne^{22}$ stand for the three neon isotopes, while Ne refers to their natural mixture.

Thomson's original crude apparatus has been improved by Aston, A. J. Dempster, and K. T. Bainbridge, and the modern *mass spectrograph*, as these instruments are called, can determine the relative masses of isotopes with great accuracy.

Further studies have shown that practically every element represents a mixture of several isotopes. While in some cases (as in gold and iodine) one isotope accounts for 100 percent of the material, in many other cases (as in chlorine and zinc), different isotopes have comparable abundances. The isotopic composition of some of the chemical elements is shown in Table 2-1.

TABLE 2-1

| Atomic Number | Name | Isotopic composition with percentage shown in parenthesis |
|---|---|---|
| 1 | Hydrogen | 1(99.985); 2(0.015) |
| 6 | Carbon | 12(98.9); 13(1.1) |
| 7 | Nitrogen | 14(99.64); 15(0.36) |
| 8 | Oxygen | 16(99.76); 17(0.04); 18(0.20) |
| 17 | Chlorine | 35(75.4); 37(24.6) |
| 30 | Zinc | 64(48.89); 66(27.81); 67(4.07); 68(18.61); 70(0.62) |
| 48 | Cadmium | 106(1.215); 108(0.875); 110(12.39); 111(12.75); 112(24.07); 113(12.26); 114(28.86); 116(7.58) |
| 80 | Mercury | 196(0.15); 198(10.02); 199(16.84); 200(23.13); 201(13.21); 202(28.80); 204(6.85) |

The remarkable fact that we notice from this table is that *whereas the atomic weight of chemical elements is not necessarily an integral number, the weight of individual isotopes is always very close to an integer.* This fact bolstered an important hypotheses, proposed a century ago by the British chemist William Prout, who considered all elements to be some kind of condensation of a single primary element: hydrogen. Prout's hypothesis, proposed very early in the development of scientific chemistry and based on the assumption of the unity of matter borrowed from medieval alchemy, was rejected by his contemporaries, who argued that the atomic weights of chlorine and mercury are far from being integral. Only after Aston's discovery of isotopes was Prout's idea reinstated in its own right, and it became, in a somewhat modified form, one of the cornerstones of the modern theory of the internal structure of matter.

## Thomson's Atomic Model

On the basis of his experiments, J. J. Thomson proposed a mode of internal atomic structure (Figure 2-6) according to which atoms consisted of a positively charged substance (positive electric fluid) distributed uniformly over the entire body of the atom, with negative electrons imbedded in this continuous positive charge like seeds in a watermelon. Since electrons repel each other but are, on the other hand, attracted to the center of the positive charge, they were supposed to assume certain stable positions inside the body of the atom. If this distribution were disturbed by some external force, such as, for example, a violent collision between two atoms in a hot gas, the electrons were supposed to start vibrating around their equilibrium positions, emitting light waves of corresponding frequencies.

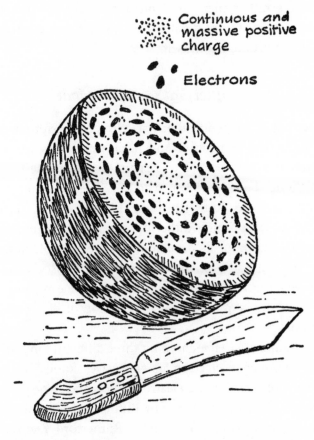

*Figure 2-6. J. J. Thomson's "watermelon" model of an atom.*

Many calculations were made in an attempt to correlate the emission frequencies of electrons in Thomson's atom with the actually observed frequencies of light emitted by different elements, but there was no success. After a number of futile efforts, it became rather clear that although Thomson's model considered an atom to be a complex system formed by positive and negative electric charges rather than an elementary indivisible body and represented a considerable progress toward the truth, it was not yet the truth itself.

## Rutherford's Atomic Model

The honor of giving the first correct description of the distribution of positive and negative charges within the atom belongs to a New Zealand-born physicist, Ernest Rutherford (1871-1937), who was later elevated to the rank of Lord Rutherford for his important scientific achievements. Young Rutherford entered physics during that crucial period of its development when the phenomenon of natural radioactivity had just been discovered, and he was the first to realize that radioactive phenomenon represent a spontaneous disintegration of heavy unstable atoms.

Radioactive elements emit three different kinds of rays: high-frequency electromagnetic waves known as $\gamma$-*rays*, beams of fast-moving electrons known as $\beta$-*rays*, and the $\alpha$-*rays*, which were shown by Rutherford to be streams of very fast-moving helium ions. Rutherford realized that very important information about the inner structure of atoms can be obtained by the study of violent collisions between onrushing $\alpha$-particles and the atoms of various materials forming the target. This started him on a series of epoch-making atomic bombardment experiments that revealed the true nature of the atom and led ultimately to

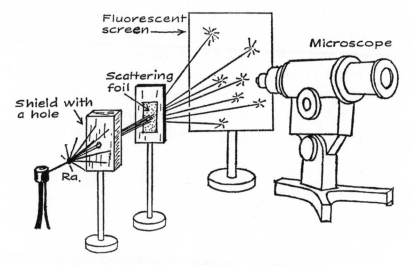

Figure 2-7. *The arrangement used by Rutherford in his "atomic bombardment" experiments.*

the present atomic energy developments. The experimental arrangement used by Rutherford in his studies were exceedingly simple (Figure 2-7): a speck of α-emitting radioactive material at *A*, a lead diaphragm, *B*, that cuts out a thin beam of α-rays, the material under investigation in the form of a piece of thin foil, *C*, a fluorescent screen *D*, and a microscope, *E*, to observe the tiny flashes of light, or scintillations, originating when an α-particle hits the screen. Before the material to be studied was inserted between the diaphragm and the fluorescent screen, scintillations were observed only in a small, sharply defined area immediately opposite the opening of the diaphragm. The introduction of the foil into the path of the α-rays, however, caused a considerable scattering of the original beam with many of the α-particles being deflected by quite large angles, and some of them even being thrown almost directly backward. Counting, through a microscope, the number of scintillations observed in different directions with respect to the original beam, Rutherford was able to construct a curve giving the relative scattering intensity as a function of the angle. One of these curves, pertaining to the scattering of radium α-particles in aluminum, is shown in Figure 2-8.

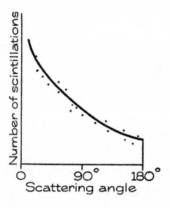

Figure 2-8. The number of scattered α-particles depends on the angle of scattering in the case of α-rays in aluminum.

Comparing the results of these experiments with the scattering that was theoretically expected on the basis of J. J. Thomson's atomic model, Rutherford noticed at once that something was drastically wrong. In fact, if the positive charge, and most of the atomic mass associated with it, were uniformly distributed through the entire volume of the atom, the collisions between the α-particles of the beam and the atoms of the target could not possibly deflect the incident particles by more than just a few degrees. In order to produce a sufficiently strong electrostatic repulsion between the positive charge of the bombarded atom and the positive charge of the incident α-particle, *all positive charges, along with most of the atomic mass, had to be concentrated in a very small central region of the colliding particles*, a region which Rutherford named the *atomic nucleus* (Figure 2-9). But, if all the positive charge of an atom is concentrated in its very center, the main body of the atom must be formed by nothing more than a swarm of negatively charged electrons moving freely through space. In order not to fall into the central nucleus under the action of the forces of electrostatic attraction, the electrons must be rotating very rapidly around the center of the system. Thus, in one bold stroke Rutherford transformed the static "watermelon model" of J. J. Thomson into a dynamic "planetary model" in which the nucleus plays role of the sun and the electrons correspond to the individual planets of the solar system.

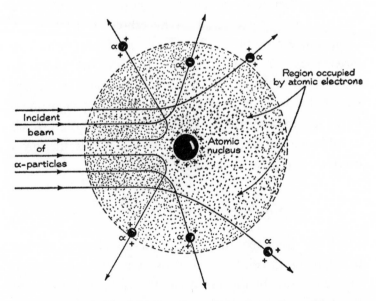

*Figure 2-9. The deflection of α-particles due to the positive charge of the atomic nucleus.*

The strong concentration of atomic mass and positive charge in the very center of the atom not only made possible the explanation of the large scattering angles observed by Rutherford in his experiments but also led to a simple quantitative expression for the number of $\alpha$-particles scattered in different directions. Since the electrons forming the outer body of the atom are very much lighter than the central nucleus and are also much lighter than the alpha particle being deflected, the role of the electrons in a collision between an atom and an alpha particle can be completely disregarded. The scattering problem is thus reduced to a collision between two mass points that repel each other with a force inversely proportional to the square of the distance. In the case of attractive forces, such as we have when planets rotate around the sun and electrons rotate around the atomic nucleus, the inverse square law of interaction leads to elliptical motion, but repulsive inverse square forces lead to hyperbolic trajectories. A comparatively simple mathematical calculation (which is, nevertheless, too complicated to reproduce here) leads to the conclusion that the number of incident particles that are scattered at a certain angle $\theta$ must be inversely proportional to the fourth power of the sine of $1/2\theta$, a conclusion which stood in perfect agreement with the results of Rutherford's original experiments shown in Figure 2-8.

## Conduction of Electricity in Solids

We have discussed the passage of an electric current through liquid solutions of acids and salts and through rarefied gases. In the first case, the current was due to the motion of positively and negatively charged ions, such as $Ag^+$ and $NO_3^-$, shouldering their way through the crowd of water molecules. In the second case, we dealt with positively charged ions flying in one direction and free negative charges, or electrons, flying in the opposite direction. But what happens when an electric current passes through solids, and why are some solids (all of them classed as metals) rather good conductors of electricity while the rest of them, known as *insulators*, hardly pass any electric current at all? Since in solid materials all atoms and molecules are rigidly held in fixed positions and cannot move freely as they do in gaseous or liquid materials, the passage of electricity through solids cannot be due to the motion of charged atoms or atomic groups. Thus, the only active electric carrier can be an electron, which, being much smaller than the atoms and molecules forming the crystalline lattice of a solid, should be able to pass between big atoms as easily as a small speedboat can pass through a heavily crowded anchorage of bulky merchantmen. Indeed, this is exactly what takes place in metallic conductors. The high electrical conductivity of the substances is inseparably connected with the presence of a large number of free mobile electrons that rush to and fro through the rigid crystalline lattices (Figure 2-10). In metals the atoms are packed considerably tighter than in other substances, and this, among other things, accounts for the relatively high density of metals. As the result of such close packing and squeezing of metallic atoms, some of the structural electrons (about one electron per atom) get detached from the main atomic body and travel at random through the metallic crystal lattice.

In the case of non-metals, such as sulfur, each atom holds tightly all of its 16 electrons, and applications of an electric field can cause nothing more than a slight deformation (electric polarization) of the atoms forming the crystal lattice. On the other hand, in the case of aluminum, only 12 out of its 13 electrons are retained in each atom while the thirteenth "black sheep" electron is detached from the basic structure and is free to move wherever the applied electric potential urges it to go.

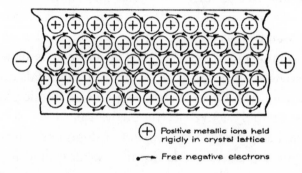

Figure 2-10. The motion of free electrons explains the passage of electric current through metals.

## Electrical Conductivity vs. Heat Conductivity

In considering the electrical conductivity of different metals we find a significant parallelism between electrical and heat conductivities; electrical conductivities of different metals are directly proportional to their heat conductivities. This fact clearly indicates that the two phenomena are closely related to each other, and, indeed, the electron theory of metals ascribes their heat conduction to the *diffusion of free electrons* from the heated end of a metallic object to the cooler end. Since the electrons in a metal can move between the atoms forming its lattice with the greatest of ease, the increased thermal agitation existing at the heated end of a metallic object spreads out very quickly toward its cooler end, in contrast to the case of insulators where all electrons are bound to their atoms and thermal agitation propagates through the material only via the interactions between neighboring vibrating molecules.

The electron theory of metals leads to a rather simple mathematical formula for the coefficients of electrical and thermal conductivities. The formula expresses these coefficients through the mass and charge of electrons, their velocity within the metal, and the distance they travel between two collisions with the atoms forming the lattice (the so-called mean free path). It turns out first of all that, for a given temperature, the coefficients of both electrical and thermal conductivities must be proportional to the number of free electrons in the metal in question, from which it follows that *the ratio of thermal and electrical conductivities must be the same for different metals at a given temperature.* Theoretical studies of the motion of free electrons through metals lead further to the conclusion that *the ratio of thermal and electrical conductivities must increase in direct proportion to the absolute temperature of the conductor.*

This statement concerning the relation between the thermal and electrical conductivities of metals constitutes the so-called Wiedeman-Franz law, which was found empirically long before the electron theory of metals was formulated. Table 2-2 shows how well this law holds for different metals at widely different temperatures.

The expected numerical value of this ratio calculated from the electron theory of metals turns out to be $2.7 \times 10^{-13}$, in good agreement with the empirical values listed in the table. The agreement between the observed and the theoretically predicted correlation between the thermal and electrical conductivities of metals and the absolute temperature is a typical example of how theoretical assumptions about the internal structure of matter increase our understanding of empirically established relations between several, at first sight, unrelated phenomena.

TABLE 2-2 THE RATIOS OF THERMAL AND ELECTRICAL CONDUCTIVITIES FOR DIFFERENT METALS DIVIDED BY THE CORRESPONDING ABSOLUTE TEMPERATURES

*(All numbers given in the table have to be multiplied by $10^{-13}$)*

| Temp | Copper | Lead | Silver | Tin | Zinc |
|---|---|---|---|---|---|
| −100°C (173° abs.) | 2.39 | 2.61 | 2.52 | 2.76 | 2.63 |
| 0°C (273° abs.) | 2.53 | 2.78 | 2.56 | 2.74 | 2.70 |
| 100°C (373° abs.) | 2.55 | 2.76 | 2.61 | 2.74 | 2.56 |

## Semiconductors

Some materials cannot be classified as either insulators or good conductors; thermal agitation of the atoms can knock loose a few electrons and permit the material to be slightly conductive. Such materials are known as *semiconductors*. A small amount of the proper kind of impurity in the crystalline structure of a semiconductor may, however, make it enormously more conductive. The three pictures in Figure 2-11 explain how and why the presence of foreign atoms in the originally completely regular lattice may lead to such a large increase of electrical conductivity.

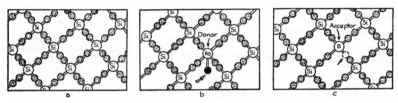

*Figure 2-11. The impurities in the crystalline structure of a semiconductor make the semiconductor very conductive.*

In Figure 2-11*a* we see a pure silicon crystal in which each atom of silicon, having a chemical valence 4, is connected with four of its neighbors by four bonds. Diagram 14*b* shows the situation that arises when one atom of silicon is replaced by an atom of arsenic, which has a valence of 5. The four valence electrons of the As atom form connections (bonds) with the four neighboring Si atoms, while the fifth "black sheep" electron is left unemployed and free to travel from place to place. The impurity atoms that give rise to free electrons in this way are known as *donors*. A reverse situation occurs when the Si atom is replaced by a trivalent atom of boron (c). In this case there will be a vacant place, or an *electron hole*, that breaks up the spotless regularity of the silicon crystal lattice. The impurity atoms that give rise to such "holes" are known as *acceptors*. A hole formed near a foreign atom present in the lattice may be filled up by an electron originally belonging to one of the neighboring silicon atoms, but in filling this hole the electron will leave a hole at the place where it was originally located. If this hole is filled by another neighboring electron, a new hole will move one step farther out (Figure 2-12). Thus, we can visualize the hole of that type as an "object" that is moving through the crystal, carrying a deficiency of negative charge, or, what is the same, a positive electric charge. Semiconductors that contain donor atoms and free electrons are known as *n-type* semiconductors; those with acceptor atoms and holes are called *p-type* semiconductors (*n* and *p* stand for a negative and positive charge of electric carriers). The electrical conductivity of n-type semiconductors is determined by the number of free electrons per unit valence and the ease with which they move through the crystal lattice, while in the case of p-type semiconductors it depends on the number and mobility of the holes.

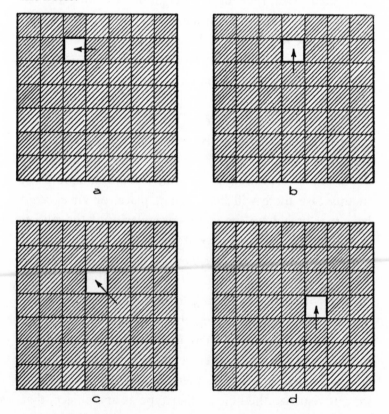

*Figure 2-12. The successive fillings of a "hole" by neighboring electrons (represented by shaded squares) make the "hole" move to the right and downwards.*

## Crystal Rectifiers

Suppose now that we put into contact two crystals: an n-type crystal containing free electrons and a p-type crystal containing electron holes (Figure 2-13). Some of the electrons from the n-region will diffuse into the p-region, while some holes from the p-region will diffuse into the n-region.

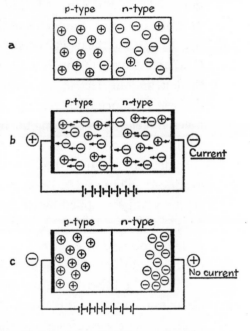

*Figure 2-13. The motion of electrons and holes across a p-n-junction. (a) In the absence of an electric field, some electrons get into the p-type crystal and some holes into the n-type crystal. (b) If the field is directed from the p-type to the n-type crystal, a continuous electric current will flow through the junction. (c) If the direction of the field is reversed, no current will flow.*

Thus the n-type crystal will become slightly positively charged whereas the p-type crystal will carry an equal negative charge. Between these opposite charges on both sides of the interface (known as an *n-p-junction*) there will be an electric force of attraction which will prevent further diffusion, and the situation will be stabilized with a certain number of holes in the n-type crystal and an equal number of electrons in the p-type crystal. It must be remembered, however, that when free electrons and electron holes exist side by side in a given material, they can be mutually "annihilated" by a free electron filling a hole. In order to compensate for the losses due to this annihilation process, a small number of electrons and holes will continue to diffuse in opposite directions through the n-p-junction.

Let us see what happens now if we apply an electric potential at the two ends of our crystal pair. If the positive pole of a battery is connected to the p-type crystal and the negative pole to the n-type crystal (Figure 2-13b), there will be a force driving the holes to the right and the electrons to the left, and an electric current will begin to flow through the system. Since both crystals are now being invaded by holes and electrons crossing the border, the rate of mutual annihilation on both sides of the n-p-junction will increase considerably, and more holes and electrons will have to be produced on both sides. These new electrons for the n-type crystal will be supplied by electrons pouring through the wire from the negative pole of the battery, while new holes will be produced by electrons leaving the p-type crystal on their way to the positive pole of the battery.

If, on the other hand, we reverse the direction of the electric potential, the situation will be quite different (Figure 2-13c). Now the electrons and the holes will be pulled in opposite directions, leaving a "no-man's land" at the n-p-junction. It is clear that under these conditions no current can flow through our double crystal. Thus we see that our device will conduct electric current in one direction but not in the opposite one. This property of one-way electric conductivity of n-p-junctions permits us to use pairs of n-type and p-type crystals for rectifying alternating current instead of the more complicated vacuum tubes.

## *Transistors*

A thin layer of a p-type crystal sandwiched between two n-type crystals can be made to function in the same manner as a vacuum tube, and is called a *transistor*. The principal advantage of transistors over vacuum tubes lies in the fact that the controlled flow of electrons takes place entirely *within solid material*. Thus it is not necessary to use a large amount of power to keep a filament red hot to "boil" electrons off into space. This, in addition to their simplicity, sturdiness, and small size, is rapidly causing transistors to take the place of vacuum tubes in many fields of electronics.

## *Solar and Radioactive Batteries*

The properties of the n-p-junction between two crystals can also be used for the direct transformation into electric energy of both solar radiation and the rays emitted by radioactive materials. When radiation is absorbed in the material of a semiconducting crystal, it knocks off some electrons from the atoms to which they belong, thus increasing the number of free electrons and electron holes. This increased number of electric carriers disturbs the electrostatic balance at the interface between the n- and p-type crystals and causes an electric current to run from the crystal containing acceptors to the crystal containing donors. A workable solar battery of this kind was recently developed in the laboratories of the Bell Telephone Company. It consists of a silicon crystal with a slight arsenical contamination (donor) through its entire body, except for a thin upper p-type layer (one ten-thousandth of an inch thick), which is contaminated by boron and serves as an acceptor. The sun's rays that fall on the upper surface of this device are absorbed in the material of the crystal, produce extra electrons and extra electron holes, and stimulate an electron potential of about one-half volt. This device has about a 20 percent efficiency, as compared with only a few percent efficiency of all previously proposed devices, and it produces a power of about 0.01 watt per $cm^2$ of its surface. A battery with a working surface of 10 sq m (about 100 sq ft) installed on the roof of a house will produce a power of 100 watts, which, when stored in ordinary electric storage batteries, is sufficient to operate a 100-watt electric bulb at night for the same number of hours that the sun was shining during the day. Because of the present high cost of producing the elements of a solar battery, it would be highly irrational to use it for the purpose of saving on the electric bill, but such batteries will undoubtedly find many useful applications, one of which has been the production of power for running the electrical equipment in experimental satellites.

The principle of the solar battery can be used also for the direct transformation of $\alpha$–, $\beta$–, and $\gamma$-rays emitted by radioactive materials, such as fission products, into the energy of electric current. If such a device can be constructed with an efficiency comparable to that of the solar battery, the fission products that result from the operation of plutonium-producing piles and various nuclear power reactors could be used to run small household gadgets and devices employed in many other walks of life.

# CHAPTER 3

# The Quantum of Radiant Energy

## The Ultraviolet Catastrophe

The radiation emitted by heated bodies represents a mixture of all different wave lengths. An increase in temperature ($T_{abs.}$) results in a rapid increase in the total amount of emitted radian energy (proportional to $T_{abs.}^4$) and in the shortening of the prevailing wave length (proportional to $T_{abs.}^{-1}$). Comparing the curves showing the distribution of energy at different wave lengths of radiation for various temperatures with the curves showing the distribution of energy (or velocities) in the molecules of a gas, we cannot help noticing a certain analogy between them: in both cases, the curves show a well-defined maximum which shifts its position with the change of temperature.

During the last decade of the nineteenth century, a British physicist and astronomer, Sir James Jeans (1877-1946), made an attempt to treat the problem of the distribution of energy between different wave lengths of radiant energy in the same statistical way as Maxwell had done in the case of the distribution of energy between different molecules of a gas. To do this, Jeans considered radiant energy of different wave lengths enclosed in a cube, the walls of which are made of ideal mirrors reflecting a full 100 percent of any radiation falling on them. Of course, this so-called Jeans's cube is just an abstraction (since there are no such mirrors) and can be used only for the purpose of purely theoretical arguments; but in physics, we very often use idealized models of this sort.

In Figure 3-1, we give a schematic picture of Jeans's cube and various waves that can exist within it. The situation is similar to that of the sound waves that can exist inside a cubical enclosure with perfectly reflecting walls, or to that of standing waves of any kind. The reflecting walls must be nodes of the standing waves, so that $l$, the distance between the walls of the cube, is an integral number of half wave lengths.

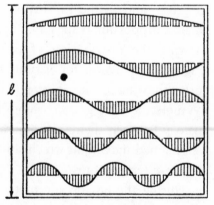

*Figure 3-1. A cross section of "Jeans's cube," showing the different waves that can exist inside it. Only the waves propagating in a horizontal direction are shown here. The black dot is a tiny coal dust particle.*

The longest wave has a wave length twice the length of the side $l$ of the cube, and the next possible wave lengths are: 1, $2/3$, $1/2$, $2/5$, etc., of $l$. We also assume that the box contains one or more "coal dust particles" that are introduced here to permit the exchange of energy between the different modes of vibrations existing in the box. (These particles are supposed to absorb the energy from the vibrations of one wave length and to re-emit it at a different wave length.)

We can now draw an analogy between the different vibrations within the Jeans's cube and the molecules of gas contained in a similar cubical enclosure. Just as in the case of the gas where the total available kinetic energy can be distributed in various ways between individual molecules, the total available radiant energy within Jeans's cube can be distributed in various ways between the vibrations of different wave lengths. The law of equipartition of energy states that *all the molecules of gas share equally in the distribution of the total available energy*, so that the mean energy of each molecule is simply the conclusion that the total radiant energy in the Jeans's cube should be equally distributed between the vibrations of all different wave lengths. But here came a very serious difficulty! Whereas the number of molecules forming a gas, though very large, is still finite, *the number of possible vibrations in Jeans's box is infinite*, since we can continue beyond any limit the sequence of possible wave lengths given above. Thus, if the equipartition law holds in this case, as it certainly should, each individual vibration would get an infinitely small share of the total energy. Since, on the other hand, the sequence of wave lengths continues indefinitely in the direction of shorter and shorter wave lengths, *all the available energy will be concentrated in the region of infinitely short waves*. Thus, if we fill Jean's cube with red light, it should rapidly become violet, ultraviolet, then turn into X-rays, into gamma rays (such as are emitted by radioactive substances), and so on beyond any limit. What happens to radiant energy in the idealized case of Jeans's cube must also hold for the radiation in all practical cases, and the light emitted by red-hot pieces of coal in the fireplace should be turned into deadly gamma rays even before it leaves the grate! Or, at least, that is what would happen if the laws of classical physics were applicable to radiant energy. This "Jeans's paradox," also known as the "ultraviolet catastrophe," gave a terrible blow to the self-satisfied classical physics of the nineteenth century and catapulted it into an entirely new field of thought and experience—now known as *quantum theory*—unprecedented in the history of physics. Although the advanced mathematics needed for detailed quantitative study of the quantum theory is not easy, the underlying concepts are not too difficult, even without extensive mathematics; and before finishing the chapter, the reader should acquire a general idea of what it is all about.

## *The Birth of the Energy Quantum*

Just before the close of the last century, in Christmas week, 1899, at a meeting of the German Physical Society in Berlin, the German physicist, Max Planck (1858-1947), presented his views on how to save the world from the perils of Jeans's ultraviolet catastrophe. His proposal was as paradoxical as Jeans's paradox itself, but it certainly was helpful. In a way, Planck's proposal can be considered as the extension of Democritus' hypothesis concerning the atomic structure of matter to the problem of radiant energy. Following Democritus, who insisted that matter cannot be subdivided into arbitrarily small portions and that one atom is the smallest possible amount of matter,

Max Planck assumed that *there must exist a smallest portion of energy,* and he gave these smallest portions of energy the name *energy quanta.* According to this revolutionary view, the bright days of the sun that pour through the windows or the soft light that radiates from a table lamp *do not represent a continuous flow of light waves, but rather a stream of individual "energy packages" or "light quanta"* (Figure 3-2). To each kind of radiation corresponds

*Figure 3-2. The old and the new picture of sunlight coming through the window: (a) old view of light beam as formed by continuous wave trains,* the amplitude *of which increases with the intensity of light; (b) new view of light beam as formed by individual vibrating "light quanta,"* the number *of which determines the intensity of light.*

a definite amount of energy which can be carried in one package, and it is just as nonsensical to talk about three-quarters of a quantum of green light as it is to talk about three-quarters of an atom of copper. Planck assumed that the light quanta of different types of radiation carry different amounts of energy and *that the amount of energy of a light quantum is inversely proportional to the wave length of the radiation, or (what is the same) directly proportional to its frequency.* Writing *v* (vibrations/second) for the frequency of the radiation and *E* for the energy of the light quantum, we can express Planck's assumption in the form:

$$E = h \times v$$

where *h* is the coefficient of proportionality known as *Planck's constant* or the *quantum constant.*

How does Planck's assumption of light quanta help to remove the perils of Jean's ultraviolet catastrophe? To understand this, let us look further into the consequences of the basic assumption that $E = hv$, that is, that radiant energy, such as light, flies about in packets of energy, the sizes of which are proportional to the frequency of the radiation. The long wave-length waves of radio have low frequencies (recall that frequency equals velocity divided by wave length; $v = v/\lambda$; and that $v = c$ for all electromagnetic radiation); hence their quanta of energy are small. Visible light, with frequencies a billion times greater, comes in quanta whose energy is also a billion times greater. Energy must be absorbed and emitted in whole quanta, exactly— no fractional parts of quanta are allowed.

The difference in the size of the demands between long-wave (low-frequency) and short-wave (high-frequency) radiation has an important effect on the application of the equipartition principle. If, for example, $6.00 must be distributed among six persons, none of whom presents any minimum demand, the fairest distribution would be to give $1.00 to each of them. Suppose, however, that Mr. A would take no less than $1.00, Mr. B no less than $2.00, and so on, up to Mr. F, who would accept no less than $6.00. It would certainly be unfair to give all six dollars to Mr. F and deprive everyone else of any share. It would not be fair either to give $5.00 to Mr. E (his minimum demand) and the remaining $1.00 to Mr. A. Clearly, the most reasonable distribution of the total money available would be to give $1.00 to Mr. A, $2.00 to Mr. B, $3.00 to Mr. C, and to deprive the Messrs. D, E, and F of any share because of their unreasonably high demands.

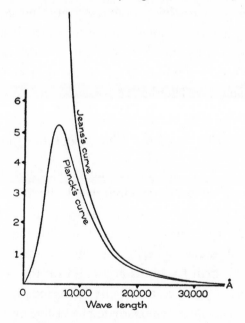

*Figure 3-3. Jeans's curve and Planck's curve representing the dependence of energy on the wave length.*

Planck did something very like this in the problem of distributing the total energy among vibrations of different frequencies existing within an enclosure of given size. His solution can be considered to take the form of a probability distribution—the vibrations with high demands have a very small chance of having their demands satisfied. Low-frequency radiation, which asks but little, has a very good chance of getting it. In other words, both ends of Planck's energy distribution curve (Figure 3-3) approach zero; at the short wave-length, high-frequency end, the radiation has practically no chance of receiving a great deal; and at the long wave-length end, the radiation stands a very good chance of receiving practically nothing. So instead of looking like the original Jeans's curve, the distribution curve obtained by Planck took a much more reasonable shape, with the main bulk of energy being given to certain intermediate wave lengths.

## The Puzzle of the Photoelectric Effect

A few years after Max Planck introduced the notion of light quanta in order to circumvent the difficulties of Jeans's ultraviolet catastrophe, a new and, in a way, much more persuasive argument for the existence of these packages of radiant energy was put forward by Albert Einstein. Einstein based his argument on the laws of the *photoelectric effect*, i.e., the ability of various materials to emit free electrons when irradiated by visible or ultraviolet light. An elementary arrangement for the demonstration of the photoelectric effect is shown in Figure 3-4. A freshly sandpapered piece of zinc, *P*, is attached to an electroscope and given a *negative* charge. If light from an electric arc is allowed to fall directly on the zinc plate, the electroscope leaves will come together, showing that the plate has lost its charge. The closer the arc light is to the plate, the more rapidly will the charge be lost; conversely, as the experiment is repeated with the arc removed to greater and greater distances, the charge will be lost more slowly. However, we find that if a sheet of ordinary glass is put between the arc light and the zinc, the zinc will retain its negative charge, even if the arc is brought very close. Also, we find that if the zinc is originally given a *positive* charge, the arc light will have little apparent effect on the rate at which the charge is lost.

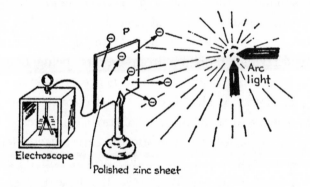

*Figure 3-4. When struck by ultraviolet radiation from an arc light, a zinc plate loses electrons.*

All this experimental evidence, said Einstein, could be quite satisfactorily explained by Planck's new idea of energy quanta. A certain amount of energy, $W_{Zn}$, is required to pull an electron loose from the attraction of the atoms in a zinc plate. According to the old classical theories, the energy of light or ultraviolet radiation spread out in spherical waves, so the amount of energy an electron could absorb from one tiny spot on such a spreading wave front would be negligible. However, Einstein argued, the old classical picture does not represent what actually happens. The entire energy of the quantum is absorbed in one bite by a single electron. Planck's relationship, $E=hv$, tells how much energy there will be in a quantum of any given frequency $v$.

For the zinc plate, $W_{Zn}$, is greater than the energy associated with a quantum of visible light, so that no matter how much visible light shines on the plate, no electron will receive enough energy to break loose. This is the situation with a sheet of glass screening the arc light—ordinary window glass shuts out the invisible but highly energetic ultraviolet radiation. With the glass removed, the ultraviolet radiation from the arc, being of higher frequency than visible light and hence of proportionally higher energy, is absorbed by the electrons in the plate. The energy of a quantum of ultraviolet is greater than $W_{Zn}$, so the electrons can escape, carrying any leftover excess energy with them in the form of kinetic energy. With the plate negatively charged, the departing electrons are repelled and the plate gradually loses its charge. A positively charged plate, however, will attract the electrons back as quickly as they escape, so there is in this case no loss of charge.

For a general energy relationship, we need consider the terms: $hv$, which is the entire energy of the quantum absorbed by the electron; $W$, the energy required to pull the electron free from the surface; and $\frac{1}{2}mv^2$, the kinetic energy of the electron as it leaves. Simple consideration of the conservation of energy gives us:

$$hv - W = \frac{1}{2}mv^2$$

For an electron to be pulled off at all, without anything left over as kinetic energy, the $hv$ of the quantum must equal $W$, which has different values for different materials. The threshold wave lengths and frequencies at which $hv$ just equals $W$ for the following three elements are:

Platinum   $\lambda = 1{,}980$ Å, or $v = 1.51 \times 10^{15}$/sec
Silver      $\lambda = 2{,}640$ Å, or $v = 1.13 \times 10^{15}$/sec
Potassium  $\lambda = 7{,}100$ Å, or $v = 4.22 \times 10^{14}$/sec

In his classical paper on this subject, published in 1905, Einstein indicated that the observed laws of the photoelectric effect can be understood if, following the original proposal of Max Planck, one assumes that *light propagates through space in the form of individual energy packages and that, on encountering an electron, such a light quantum communicates to the electron its entire energy.*

This revolutionary assumption explains quite naturally the observed fact that the increase of the intensity of light leads to the increase of the number of photoelectrons, but not of their energy. More intense light means that more light quanta of the same kind will fall on the surface per second, and, since a single light quantum can eject one and only one electron, the number of electrons must increase correspondingly. On the other hand, by decreasing the wave length of incident light we increase the frequency and, consequently, the amount of energy carried by each individual light quantum, so that in each collision with a free electron in the metal, these quanta will communicate to it a correspondingly larger amount of kinetic energy.

## *The Compton Effect*

The Planck-Einstein picture of individual energy packages, or light quanta, forming a beam of light and colliding with the electrons within matter intrigued the mind of an American physicist, Arthur Compton, who, being of a very realistic disposition, liked to visualize collisions between light quanta and electrons as similar to those between ivory balls on a billiard table. He argued that, in spite of the fact that the electrons forming the planetary system of an atom are bound to the central nucleus by attractive electric forces, these electrons would behave exactly as if they were completely free if the light quanta which hit them carry sufficiently large amounts of energy. Suppose that a black ball (electron) is resting on a billiard table (Figure 3-5) and is bound by a string to a nail driven into the table's surface and that a player, who does not see the string, is trying to put it into the corner pocket by hitting it with a white ball (light quantum). If the payer sends his ball with a comparatively small velocity, the string will hold during the impact and nothing will come of this attempt. If the white ball moves somewhat faster, the string may break, but in doing so it will cause enough disturbance to send the black ball in a completely wrong direction. If, however, the kinetic energy of the white ball exceeds, by a large factor, the work necessary to break the string that holds the black ball, the presence of the string will make practically no difference, and the result of the collision between the two balls will be practically the same as if the black ball were completely unbound.

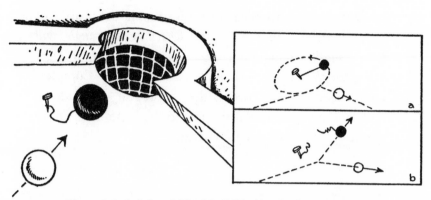

*Figure 3-5. A tied-up billiard ball (black) is hit by a moving ball (white). If the white ball moves slowly (a), the effect of the string will be essential, but if it moves fast (b), the result of the collision will be the same as if the black ball were not tied at all.*

Compton knew that the binding energy of the outer electrons in an atom is comparable to the energy of the quanta of visible light. Thus, in order to make the impact overpoweringly strong, he selected for his experiments the energy-rich quanta of high-frequency X-rays. The result of a collision between X-ray quanta and (practically) free electrons can be indeed treated very much in the same way as a collision between two billiard balls. In the case of an almost head-on collision, the black ball (electron) will be thrown at high speed in the direction of the impact, while the white ball (X-ray quantum) will lose a large fraction of its energy. In the case of a side hit, the white ball will lose less energy and will suffer a smaller deflection from its original trajectory. In the case of a mere touch, the white ball will proceed practically without deflection and will lose only a small fraction of its original energy. In the language of light quanta, this behavior means that in the process of scattering, *the quanta of X-rays deflected by large angles will have a smaller amount of energy and, consequently, a larger wave length.* The experiments carried out by Compton confirmed, in every detail, the theoretical expectations and thus gave additional support to the hypothesis of the quantum nature of radiant energy.

# CHAPTER 4

# The Bohr Atom

## *Bohr's Quantum Orbits*

When Rutherford (at that time just plain Ernest Rutherford and not yet Sir Ernest or Lord Rutherford) was at the University of Manchester performing his epoch-making experiments that demonstrated the existence of the atomic nucleus, a young Danish physicist named Niels Bohr (1886-1962) came to work with him on the theoretical aspects of the atomic structure problem. Bohr was highly impressed by Rutherford's new atomic model in which the electrons revolved around the central nucleus, in very much the same way as the planets revolve around the sun, but he could not understand how such a motion could be at all possible in an atom. The planets of the solar system are electrically neutral, but atomic electrons are heavily charged with negative electricity (in fact, there is not much more to an electron than its electric charge!). It was well known from the theory of electricity that oscillating or revolving electric charges always emit electromagnetic waves. The emission of electromagnetic waves must result in the loss of energy by the emitting particle, so that the electrons in the Rutherford model were bound to spiral toward the central nucleus and fall into it when all of their rotational energy was spent on radiation.

Bohr calculated that the emission of electromagnetic waves (which in the case of the atom corresponds to light waves of different lengths) would cause the electrons forming an atomic system to lose all their energy and fall into the nucleus with one hundred-millionth of a second! Thus, on the basis of conventional mechanics and electrodynamics, the planetary system of electrons revolving around the atomic nucleus as visualized by Rutherford could not exist for more than an extremely short period of time. This was in direct contradiction of the fact that atoms *do exist permanently* and do not show any tendency to collapse. How could it possibly be? Bohr's solution of this conflict between the conclusions of conventional mechanics and the fact of nature was straightforward and just: *Since nature cannot be wrong, conventional mechanics must be wrong, at least when applied to the motion of electrons within an atom.* In making this revolutionary statement concerning the motion of electrons within an atom, Bohr followed the precedent established by Planck and Einstein, who had some time before declared that the good old Huygens light waves were not what they were supposed to be according to the conventional views, but rather a bunch of individual oscillating light quanta.

It is always much easier to say that something is wrong than to find a way to make it right, and Bohr's criticism of conventional mechanics in the case of atomic electrons would be of no value whatsoever if he could not show a way out of the difficulty. The way he proposed was so odd and unconventional that he kept the manuscript locked in his desk for almost two years before he decided to send it in for publication. When this epoch-making paper finally appeared in 1913, it sent out a shock wave of amazement through the world of contemporary physics!

Defying the well-established laws of classical mechanics and electrodynamics, Bohr stated that in the case of the motion of electrons within an atom the following postulatory rules must strictly hold:

*I. From all the mechanically possible circular and elliptical orbits of electrons moving around the atomic nucleus, only a few highly restricted orbits are "permitted," and the selection of these "permitted" orbits is to be carried out according to specially established rules.*

*II. Circling along these orbits around the nucleus, the electrons are "prohibited" from emitting any electromagnetic waves, even though conventional electrodynamics says they should.*

*III. Electrons may "jump" from one orbit to another, in which case the energy difference between the two states of motion is emitted in the form of a single Planck-Einsteinian light quantum.*

The whole thing sounded quite incredible, but it *did* permit Bohr to interpret the regularities of spectra emitted by various atoms and to construct a consistent theory of internal atomic structure. We will limit our discussion here to the case of the hydrogen atom, which contains a single electron revolving around the nucleus. Bohr's original restrictions concerning the motion of the electron in a hydrogen atom pertained strictly to the case of circular motion and required that the angular momentum of the electron be an integral multiple of $h/2\pi$, where $h$ is Planck's constant. Bohr's assumptions also demanded that *the "permitted" orbits be only those whose radii $2^2$, $3^2$, $4^2$, $5^2$, etc., larger than a certain minimum radius:* $r_0$. The set of these "permitted" orbits is shown in Figure 4-1.

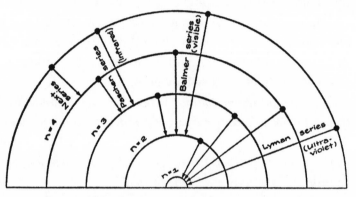

Figure 4-1. *Different series of spectral lines originating in the electron transitions in a hydrogen atom. (The radii of the orbits are not to scale.)*

Since, according to Bohr's postulate, *the radii of permitted orbits increase as the squares of the integers*, we can conclude from the regular laws of mechanics (or, rather, from what is left of these laws) that *the energy of motion along these orbits decreases as the inverse squares of the integers.*

Now, according to the second of Bohr's postulates, an electron does not emit any radiation while moving along a given orbit but does so when it "jumps" from one orbit to another. Consider, for example, the "jump" of an electron from the third orbit to the second one. Since the corresponding energies are proportional to $1/3^2 = 1/9$ and $1/2^2 = 1/4$, the energy difference liberated in this jump must be proportional to $(1/4 - 1/9)$. In the case of a "jump" taking place from the fourth, fifth, etc., orbits to the second one, the corresponding energy differences are expected to be proportional to $(1/4 - 1/16)$, $(1/4 - 1/25)$, etc.

Remembering that, according to Bohr's third postulate, the energy liberated in such a jump is transformed directly into a single light quantum and that, according to the Planck-Einstein hypothesis, the energy of a light quantum is proportional to its frequency (i.e., $E = hv$), we conclude that *the frequencies of light emitted by a hydrogen atom must be proportional to $(1/4 - 1/n^2)$ where n is an integer. But, this is exactly the "Balmer formula" for the hydrogen spectrum.* This formula states that the frequencies of the observed lines in that spectrum are exactly proportional to the difference between the inverse square of 2, i.e., $1/4$, and the inverse squares of 3, 4, 5, etc.

A question naturally arises about how the electron in the hydrogen atom could get up to a higher energy level in order to jump back and emit energy. Obviously, the electron can get into a higher energy orbit only by absorbing energy. The absorbed energy may come from collisions, if the gas is heated to a high temperature. It may come from the energy of an electric spark or cathode-ray tube discharge, or *it may arise from the gas absorbing, from radiation falling on it, those same frequencies that it is able to emit.* This last is the explanation of the dark lines crossing the spectrum of the sun. The highly compressed gases of the deep-lying solar photosphere emit a continuous spectrum including all frequencies. The atoms in the low-pressure upper atmosphere of the sun absorb from this continuum of radiation those photons whose energy is just exactly enough to raise an electron from one energy level to another. Thus the frequencies of the absorbed photons are exactly the same as the frequencies of the photons that the atom radiates.

Does this highly artificial picture of light emission by a hydrogen atom really make any sense? Were Bohr's postulates not specially adjusted so as to lead in the end to the empirically established Balmer's formula? Certainly they were! But this is exactly how a new theory is usually introduced in physics. Newton introduced the notion of universal gravity in order to interpret the observed motion of the moon around the earth and the planets around the sun, and in the very same way Bohr introduced his three postulates pertaining to electron motion in an atom and light emission by "jump" processes in order to interpret the observed laws of atomic line spectra. However, the criterion for the validity of any new theory in physics is not only that this theory should give a correct interpretation of the previous observations but that it also *predict* things which be later confirmed by direct experiment. In this respect, Bohr's theory of atomic structure came out with flying banners. The theory was constructed in order to interpret Balmer's formula, and this was achieved by ascribing the line of the Balmer series to the fact that electrons "jump" from various higher orbits to the *second orbit* in the hydrogen atom. Spectral lines corresponding to the first type of jump (i.e., to the first orbit) were expected to be located in the ultraviolet part of the spectrum and were, in fact, found there by the Harvard spectroscopist,

Lyman. The line corresponding to jumps to the third orbit lie in the infrared region where they were actually found by the German spectrocopist, Daschen. The fact that Bohr's atomic model, constructed especially in order to explain the Balmer series alone, leads to further conclusions that were later verified by experiment makes it a *really good theory*.

## Production of X-rays

Corresponding to the single electron of the hydrogen atom, the outermost electrons of other atoms also have orbits of higher energy to which they can jump when excited by heat or strong electrical fields, or by the absorption of radiation of the proper frequency and energy. In dropping back to their normal levels, these electrons radiate frequencies characteristic of the atoms to which they belong. As in the hydrogen atom, these outer-electron frequencies are in the range of visible light, infrared, or ultraviolet. X-rays are likewise produced by electron jumps from one energy level to another, but the energy differences radiated away are enormously greater than those associated with outer electrons, and the resulting X-rays are of very high frequency.

An X-ray tube is similar in principle to a cathode-ray tube, and electrons emitted from the cathode are accelerated through a vacuum by a potential of many thousands of volts to strike against a metal anode. This bombardment is so energetic that electrons are knocked out of the inner shells of the atoms of the anode. When a heavy metal atom thus loses an inner electron, the vacancy is filled by one of the outer electrons falling down to take its place, and the energy differences between inner and outer electron shells are enormous, particularly for the large atoms of heavy metals. These large energy differences, by the $E = hv$ relationship, are radiated away as energetic photons of very high frequency.

## Elliptical Quantum Orbits

Bohr's paper in which the notion of quantum orbits of atomic electrons was first introduced caused a deluge of publications all over the world, and within a few years the quantum theory of atomic structure developed into one of the most important branches of physics. The first step in this development was the generalization of Bohr's idea of circular quantum orbits. In a hydrogen atom for the case of elongated elliptical orbits. This extension of Bohr's scheme was carried out by the German physicist, A. Sommerfeld, and is illustrated in Figure 4-2. While retaining, unchanged, the first of Bohr's orbits, Sommerfeld added one elliptical orbit to Bohr's second orbit, two elliptical orbits to Bohr's third orbit, etc. Although the elliptical orbits added by Sommerfeld had different geometrical shapes, they nevertheless corresponded to almost the same energies as Bohr's circular orbits so that Bohr's original explanation of the lines of the Balmer series in hydrogen remained unchanged.

The purpose of the modification introduced by Sommerfeld was to allow more freedom in choosing the "permitted" orbits in more complicated atoms that contain more than one electron. It was known that for sodium or potassium, for example, the lines of the emission spectrum form a series very similar to the Balmer series of hydrogen, but that in the spectrum each line consists of several closely spaced components called the "multiplicity structure." Sommerfeld explained this multiplicity structure of spectral lines as owing to the fact that, in the presence of other electrons, the energy of elliptical orbits becomes slightly different from that of the circular ones, which results in a corresponding change in the frequencies of light quanta emitted during the electron jumps from one of these orbits to the other.

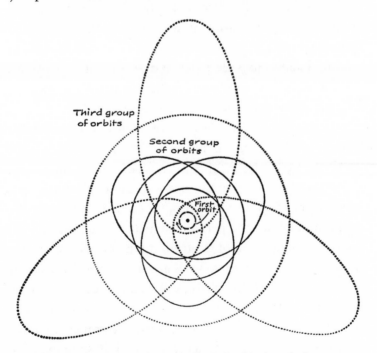

*Figure 4-2. The circular and elliptical orbits in a hydrogen atom, according to Sommerfeld.*

## *The Pauli Principle*

Since we are now acquainted with all the possible orbital motions of atomic electrons that are permitted by the Bohr-Sommerfeld rules, we can tackle the question of what the combined internal motion looks like in atoms that contain many electrons. Since atomic electrons have a tendency to jump from higher orbits to lower ones, emitting their excess energy in the form of light quanta, the normal state of any atom would be the state in which all atomic electrons move in a "ring around the rosy" along the first of Bohr's orbits. With atoms of increasing atomic number, this ring would become more and more crowded because it would have to accommodate more electrons and also because its radius would become smaller and smaller owing to the stronger electric attraction exercised by the central nucleus.

If this were true, the size of atoms would decrease rapidly with atomic number, and an atom of lead, for example, would be much smaller than one of aluminum. Experiment tells us that this is not so; although atomic volumes show  periodic variations, they remain essentially the same throughout the periodic system of the elements. To avoid this congestion of electrons on the innermost orbit, a new postulatory restriction was apparently necessary, and it was introduced by the physicist, W. Pauli. According to the Pauli principle,* *any given quantum orbit in an atom can be occupied by no more than two electrons.* Electrons are known to rotate rapidly around their axes like little spinning tops, and the Pauli principle permits two electrons to move along the same orbit only under the condition that they spin in opposite directions.

*Called the *exclusion principle* by Pauli himself.

## Electron Shells and the Periodic System

We are now in a position to find the pattern of electron motion in the atoms of the elements. The element immediately following hydrogen is helium, the atom of which contains two electrons. If these two electrons spin in opposite directions, both can be accommodated on the first (circular) Bohr's orbit, as shown in Figure 4-3a.

The next element is lithium, with three atomic electrons. Since no place is available on the first Bohrs orbit, the third electron has to be placed on the next higher energy shelf, i.e., either on the second Bohr's circular orbit or on the corresponding elliptical (Sommerfeld's) orbit. A detailed analysis of this situation indicates that there are three elliptical orbits of this type, which are identical in shape and in energy *but* are oriented in space in three different ways, and that the energy of motion along these elliptic orbits is slightly lower than that for the second circular orbit. Thus, the normal state of the lithium atom will be as shown in Figure 4-3b.

As we proceed along the natural sequence of elements, more and more electrons are placed on the second set of orbits until we reach neon, the tenth element. In neon, eight electrons are accommodated on the second shelf, and the pattern of electron motion within the atom looks as shown in Figure 4-3c. This shelf of energy or, as physicists call it, "electron shell," is completely occupied, and if there are more electrons, they must be placed on the third shelf (or shell). Thus, the atom of the eleventh element, sodium, will have two completed electron shells (with 2 and 8 electrons, respectively) and one extra electron that is to be accommodated on the third energy shelf (Figure 4-3d).

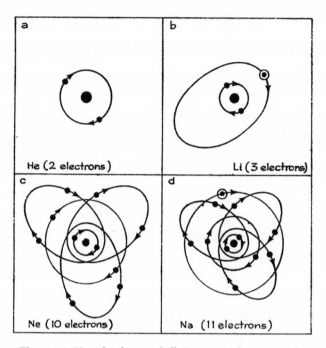

Figure 4-3. How the electron shells in more and more complex atoms are filled. As the atomic number increases, the electron shells shrink, because of the increased nuclear electric charge, so that the size of the atoms remains on the average constant. Electrons forming the beginning of a new shell are indicated with a circle around them.

An atom of sodium is, in a way, similar to an atom of lithium, since in both cases there is one extra electron moving outside of a previously completed electron shell; this similarity accounts for the similarity of their chemical properties and also of their optical spectra. It goes without saying that the prototype of sodium and lithium atoms is the hydrogen atom itself.

The element following sodium is magnesium, which has two extra electrons beyond the completed shell that give it chemical properties similar to those of beryllium. It is followed by aluminum with 3 outer electrons that make this atom chemically similar to boron, and so on, until we come to argon. Argon has 18 electrons; 2 in the first shell, 8 in the second, and 8 in the third, which is a very stable configuration. The next element is potassium, which has one extra electron beyond the outer shell of argon and is the fourth member of the H, Li, Na sequence of chemically similar elements.

We see that the combination of the Bohr-Sommerfeld notion of quantum orbits and Pauli's principle concerning the orbital cohabitance of atomic electrons leads to a simple and complete explanation of the periodic properties of the elements.

## Chemical Valence

This view of atomic shell structure gives us a simple explanation of the nature of the chemical valence of different elements. We can show, on the basis of the quantum theory, that atoms which have an almost competed shell have a tendency to take in extra electrons in order to finish this shell and that atoms which have just the beginning of a new electron shell have a tendency to get rid of these extra electrons. For example, chlorine (atomic number 17) has 2 electrons in the first shell, 8 in the second, and 7 in the third, which makes the outer shell short one electron. On the other hand, a sodium atom (atomic number 11) has 2 electrons in the first shell, 8 in the second, and only 1 electron as the beginning of the third shell.

Under these circumstances, when a chlorine atom encounters a sodium atom, it "adopts" the latter's lonely outer electron and becomes Cl–, while the sodium atom becomes Na+. The two ions are now held together by electrostatic forces and form a stable molecule of table salt. Similarly, an oxygen atom that has two electrons missing from its outer shell (atomic number = 8= 2 +6) tends to adopt two electrons from some other atom and can thus bind two monovalent atoms (H, Na, K, etc.) or one bivalent atom, such as magnesium (atomic number = 12 = 2 + 8 + 2), which has two electrons to lend. An example of chemical binding of this kind is shown in Figure 4-4. It also becomes clear why the noble gases, which have all their shells completed and have no electrons to give or to take, are chemically inert.

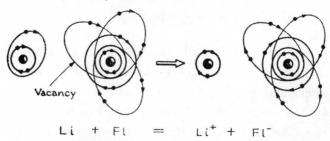

Figure 4-4. The formation of lithium fluoride.

# CHAPTER 5

# Wave Nature of Particles

## De Broglie Waves

Although Bohr's theory of atomic structure was immensely successful in explaining a large number of known facts concerning atoms and their properties, the three fundamental postulates underlying his theory remained quite inexplicable for a long period of time. The first step in the understanding of the hidden meaning of Bohr's discrete quantum orbits was made by a Frenchman, Louis de Broglie, who tried to draw an analogy between the atoms and the discrete sets of mechanical vibrations that are observed in the case of violin strings, organ pipes, etc. "Could it not be," de Broglie asked himself, "that the optical properties of atoms are due to some kind of standing waves enclosed within themselves?" As a result of these considerations, de Broglie came out with his hypothesis that *the motion of electrons within the atom is "guided" by a peculiar kind of waves which he called "pilot waves."* According to these unconventional views, each electron circling around an atomic nucleus must be considered as being accompanied by a standing wave that runs around and around the electronic orbit. If this is true, the only orbits that would be possible are those whose lengths are an integral multiple of the wave length of the corresponding de Broglie wave.

Figure 5-1 shows the application of de Broglie's idea to the first three orbits of the Bohr hydrogen atom. In order to have $n$ complete wave lengths ($\lambda_n$) fit into the circumference of the $n$th orbit, de Broglie had to assume that

$$\lambda_n = \frac{h}{mv}$$

The above relationship states de Broglie's fundamental hypothesis, that *the wave length of the wave associated with a moving particle is equal to Planck's quantum constant divided by the momentum of the particle.*

Although we have justified this by considering an electron in a Bohr atom, if electrons are really accompanied by these mysterious de Broglie waves while moving along the circular orbits within an atom, the same must be true for the free flight of electrons as observed in free electron beams. And, if the motion of electrons in the beams is "piloted" by some kind of waves, we should be able to observe the phenomena of *interference* and *diffraction* of electron beams in the same way that we observe these phenomena in beams of light. For a stream of moving electrons, we must assume for the wave length of the "pilot waves" the same formula that applied to orbital electrons within the atom:

$$\lambda = \frac{h}{mv}$$

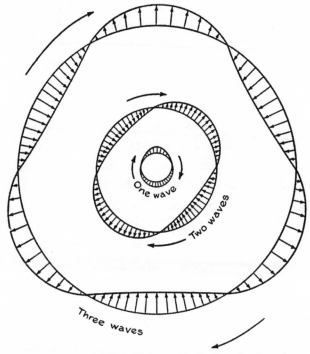

*Figure 5-1. De Broglie's waves as applied to the first three orbits of Bohr's atom.*

For the electron beams used in laboratories, this wave length comes out to be much shorter than that of ordinary visible light and is comparable, in fact, with the wave lengths of X-rays, i.e., about $10^{-8}$ cm. Thus, it would be futile to try to observe the diffraction of electron beams by using ordinary optical diffraction gratings. We should instead employ a method similar to that used in studying X-ray spectra. To examine X-ray spectra, physicists use a "crystal spectrograph" that was developed by the British physicists, W. H. and W. L. Bragg (father and son). A beam from an X-ray tube falls on the surface of a crystal and is reflected successively from the molecular layers that form the crystalline surface. Depending on the angle, the wavelets reflected from different layers may be "in phase" with each other or "out of phase," thus leading to the intensification or to the reduction of the intensity of the reflected beam.

Two American physicists, C. J. Davisson and L. H. Germer, used a similar arrangement in their experiments on electron diffraction, the only difference being that the beam of X-rays was replaced by a beam of electrons, which was accelerated by passing it though an electric field between two grids. The result of Davisson and Germer's experiment gratified their expectations, and they obtained a genuine diffraction pattern of an electron beam with the wave length corresponding exactly to the value predicted by the de Broglie theory.

A few years later, a German physicist, O. Stern, repeated the experiments of Davisson and Germer by using a molecular beam of hydrogen molecules and helium atoms (Figure 5-2) instead of an electron beam and found that the diffraction phenomenon exists in that case, too. Thus, it became quite evident that in material particles as small as atoms and electrons, the basic ideas of classical Newtonian mechanics should be radically changed by introducing the notion of "pilot waves" guiding material particles in their motion.

The once definite distinction between waves and particles seems to have broken down. There are many sorts of interference experiments in which light shows itself to be unquestionably a wave phenomenon; yet in the photoelectric effect, it concentrates all its energy on a single electron, as though it were a bullet-like particle. And now electrons and atoms, so surely particles, behave in some experiments as though they were waves.

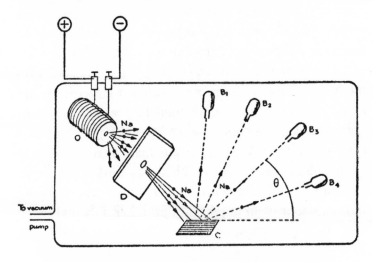

Figure 5-2. O. Stern's experiment demonstrating the diffraction of a molecular beam. A beam of sodium atoms from the oven, O, is passed through a diaphragm, D, and falls on a crystal, C. The atoms reflected from the crystal in different directions are collected in bottles, B1, B2, B3, etc., and their amounts are measured. The results show a strong maximum in the direction required by the ordinary law of reflection, and a number of secondary maxima corresponding to a diffraction pattern.

The wave aspect and the particle aspect seem to be so mutually contradictory that it is quite natural to ask which one is "really" correct for a beam of light or an electron. The modern physicist will say that neither one is "really" correct. We are trying to make the submicroscopic world of the photon and the atom fit models we imagine as being analogous to tiny bullets and tiny ripples on water tanks. The world of the atom and the photon cannot be described in the same terms we use to describe the behavior of the macroscopic world of matter-in-bulk. That we cannot is attested by the wave-particle dilemma, and the contradictions of a similar nature we run into when we try. The physicist has mathematical equations whose solutions will give the correct answers, whether it be wave or particle that is involved—but to these equations no model or picture is connected. We must either learn not to ask questions about which model is "really" correct, or not to think of any model at all. We quite plainly cannot have it both ways. An intricate mathematical method for handling this kind of problem was worked out by an Austrian physicist, E. Schroedinger, and it represents the subject matter of an important but rather difficult branch of modern theoretical physics known as *wave mechanics*.

## The Uncertainty Principle

Now things seem to be going from bad to worse. First, we had Bohr's "quantized orbits" that looked like railroad tracks along which the electrons were running around the atomic nucleus. There these tracks were replaced by mysterious "pilot waves" that were supposed to provide "guidance" for the electrons in their orbital motion. It all seemed to be against common sense, but, on the other hand, these developments of the quantum theory provided us with the most exact and most detailed explanation (or description) or the properties of atoms—their spectra, their magnetic fields, their chemical affinities, etc. How could it be? How could what was, at first sight, a nonsensical picture lead to so many positive results? Here we repeat that modern physics extends its horizons far beyond the everyday experience upon which all the common-sense ideas of classical physics were based. We are thus bound to find striking deviations from our conventional way of thinking and must be prepared to encounter facts that sound quite paradoxical to our ordinary common sense. In the case of the theory of relativity, the revolution of thought was brought about by the realization that space and time are not the independent entities they were always believed to be, but are the parts of a unified space-time continuum. In the quantum theory we encounter a non-conventional concept of *the minimum amount of energy*, which, although of no importance in the large-scale phenomena of everyday life, leads to revolutionary changes in our basic ideas concerning motion in tiny atomic mechanisms.

Let us start with a very simple example. Suppose we want to measure the temperature of a cup of coffee but all we have is a large thermometer hanging on the wall. Clearly, the thermometer will be inadequate for our purpose because when we put it into the cup it will take so much heat from the coffee that the temperature shown will be considerably less than that which we want to measure. We can get a much better result if we use a small thermometer that will show the temperature of the coffee and take only a very small fraction of its heat content. The smaller the thermometer we use for this measurement, the smaller is the disturbance caused by the measurement. In the limiting case when the thermometer is "infinitely small," the temperature of the coffee in the cup will not be affected at all by the fact that the measurement was carried out. The common-sense concept of classical physics was that this is always the case in whatever physical measurements we are carrying out, so that we can always compute the disturbing effect of whatever gadget is used for the measurement of some physical quantity and get the exact value we want. This statement certainly applies to all large-scale measurements carried out in any scientific or engineering laboratory, but it fails when we try to stretch it to such tiny mechanical systems as the electrons revolving around the nucleus of the atom. Since, according to Max Planck and his followers, energy has "atomic structure," *we cannot reduce the amount of energy involved in the measurement below one quantum*, and making exact measurements of the motion of electrons within an atom is just as impossible as measuring the temperature of a demitasse of coffee by using a bulky bathtub thermometer! But, whereas we can always get a smaller thermometer, it is absolutely impossible to get less than one quantum of energy.

A detailed analysis of the situation indicates that *the existence of the minimum portions of energy prohibits us from describing the motion of atomic particles in* the *conventional way by giving their successive positions and velocities*. Both of these quantities can be known only within certain limits, which although negligibly small for the large-scale object, become of paramount importance within tiny mechanisms. This uncertainty in the knowledge of the coordinate $x$ and the velocity $v$ of a particle can be expressed mathematically by writing $x \pm \Delta x$ and $v \pm \Delta v$, which means that all we can say is that the value of the coordinate lies somewhere between $x - \Delta x$ and $x + \Delta x$, and that the value of the velocity lies somewhere between $\upsilon - \Delta\upsilon$ and $\upsilon + \Delta\upsilon$. The German physicist, W. Heisenberg, has shown that the quantities $\Delta x$ and $\Delta\upsilon$ are subject to the relation:

$$\Delta x \times \Delta v = \frac{h}{m}$$

where $m$ is the mass of the particle and $h$ Planck's quantum constant which has the numerical value $6.77 \times 10^{-27}$. The smaller $m$ is, the more restricting is Heisenberg's uncertainty relation. If we apply it, for example, to a particle weighing 1 mg ($10^{-3}$ gm), we find that:

$$\Delta x \times \Delta v = 10^{-24}$$

which may mean that if the uncertainty of the position is $\pm 0.000000.000001$ cm, the uncertainty of the velocity is $\pm 0.000000.000001$ cm/sec. Clearly, such small uncertainties are of no importance! However, using for $m$ the mass of an electron (about $10^{-27}$ gm), we obtain:

$$\Delta x \times \Delta v = 1$$

which indicates that there may be an uncertainty of $\pm 1$ cm in the position and $\pm 1$ cm/sec in the velocity. These uncertainties are large enough to make the classical picture of the orbital motion of atomic electrons completely invalid. De Broglie's "pilot waves" give us a new way of describing the motion of atomic particles in which, instead of speaking about their trajectories, *we speak only about the probability of finding the particle in one or another location in space.* In fact, *the intensity of these waves gives us directly this probability.*

## Anti-Particles

Up until about a quarter of a century ago, physicists recognized only two kinds of elementary particles from which matter was supposedly built. They were *protons*, the relatively massive particles carrying a positive electric charge, and the much lighter negatively charged *electrons*. but this simple picture was distorted in 1929 by a British physicist, P. A. M. Dirac, who was at that time busy trying to reconcile the basic principles of the quantum theory with those of Einstein's theory of relativity. On the basis of very abstract theoretical considerations, Dirac came to the conclusion that, apart from the "ordinary" electrons which rotate around atomic nuclei or fly through vacuum tubes, there must also exist an incalculable multitude of "extraordinary" electrons, their presence escapes any possible experimental detection. The "ordinary" electrons studied by physicists and utilized by radio engineers are those few excess particles that cause an "overflow" of "Dirac's ocean," which is formed by the "extraordinary" particles, and they thus can be observed individually. If there is no such "overflow" nothing can be observed, and we call the space empty. The nearest simple analogy is that of a deep water fish who never rises to the surface of the ocean. Of course, fish in general do not possess much in the way of brains, but even if they were as intelligent as modern theoretical physicists they would find it difficult to conceive of the idea of a surrounding medium of water provided this medium is completely uniform and (as it is in the case of "Dirac's ocean") frictionless. In a similar way, Dirac's ocean surrounding us on all sides and extending into infinity in all directions remains unobservable to us. In a sense, Dirac's theory brings us back to the old-fashioned idea of the "all-penetrating world ether," but in an entirely new fashion.

In addition to having the property of not being observable by any physical means, these "extraordinary" electrons possess, according to Dirac, a "negative inertial mass," which means that when they are pushed in one direction by any physical force, they move in exactly the opposite direction. Of course, for a conventional physicist the idea of a "negative mass" seems just as nonsensical as the idea of a vacuum tightly packed by extraordinary electrons, and during the first couple of years after its publication, Dirac's paper was subjected to all kinds of criticism. The criticism stopped abruptly in 1931, however, when an American physicist, Carl Anderson, confirmed by direct observation the existence of the new particles predicted by Dirac's theory.

We have said above that, because of their uniform distribution, the "extraordinary" electrons forming in Dirac's ocean are invisible to observation, but what happens if one of these particles is absent, leaving in its place an empty "hole"? This hole in the uniform distribution of negatively charged particles represents the *lack of a negative charge,* which is equivalent to the *presence of a positive charge.* Thus, the electrical instruments used in our physical laboratories would register this "hole" as a positively charged particle with the same numerical value of charge as an ordinary electron, but with the opposite sign. The reader will recall that the notion of "holes" in the uniform distribution of electrons in semi-conductors led to a successful explanation of their properties. But, whereas in that case the notion of a "hole" can be readily visualized on the basis of an ordinary picture of the electrical nature of matter, Dirac's "holes" belong to a much more abstract physical picture.

It is also easy to see that when experimentalists study the motion of such a "hole" under the action of any external physical force, they will ascribe to it an ordinary positive mass. Returning to our intelligent deep water fish, imagine that it observes a series of air bubbles rising to the surface from a sunken submarine. Being accustomed to seeing objects in the water moving downward and sinking toward the bottom under the action of the forces of gravity, our fish would be surprised to see these silvery spheres move in the opposite direction; if our fish were intelligent enough, he might be inclined to ascribe to these unusual rising objects a "negative mass." For Dirac's ocean of "extraordinary" negative electrons possessing a negative mass, we conclude that a "hole" in this distribution must possess a mass opposite to that of the particles forming it, i.e., a positive mass. Thus, through *double negation*, we find that the *"holes" in Dirac's ocean must behave as ordinary particles carrying a positive electron charge and a positive mass.* They are called *anti-electrons, positive electrons*, or simply *positrons*.

From what has been said, we can conclude that in order to form a positron we have to remove a negative electron from its place in Dirac's ocean. But when this electron is removed from the uniform distribution of the negative electric charge, it becomes observable as an ordinary negatively charged particle. Thus, *the positive and negative electrons always must be formed in pairs.* We often call this process *the creation of an electron pair*, which is not quite correct because the pairs of electrons are not created from nothing but are formed at the expense of the energy spent in carrying out the process of their formation. According to Einstein's famous law of the equivalence of mass and energy

$$\left( E = Mc^2 \text{ or } M = \frac{E}{c^2} \right)$$

the energy necessary to produce two electron masses is equivalent to about $1.64 \times 10^{-6}$ ergs. Thus, if we irradiate matter with gamma rays of this and higher energies, we should be able to induce the formation of pairs of positive and negative electrons. The electron pairs discovered by Anderson were produced in atmospheric air, and also in metal plates placed in a detecting cloud chamber, by the high-energy gamma radiation that is associated with the cosmic rays which fall on the earth from interstellar space. Following this discovery, physicists learned to produce electron pairs by irradiating different materials by the high-energy gamma rays that are emitted by natural radioactive substances.

The opposite of the "creation" of an electron pair is the "annihilation" of a positive electron in a collision with an ordinary negative electron. According to the above-described picture, the annihilation process occurs when an ordinary negative electron, which moves "above the rim" of the completely filled Dirac's ocean, finds a "hole" in the distribution and falls into it. In this process the two individual particles disappear, giving rise to gamma radiation with a total energy equivalent to the vanished mass radiating from the place of encounter. Dirac's original theory of "holes" not only predicted the existence of positive electrons before their experimental discovery but also gave an excellent mathematical apparatus for calculating the probabilities of the formation of electron pairs under different circumstances, as well as the probability of their annihilation in casual encounter. All the predictions of this theory stand in perfect agreement with experimental evidence.

## Anti-Protons and Anti-Neutrons

Ever since the experimental confirmation of Dirac's theory of anti-electrons, physicists have been interested in finding the *anti-protons* that should be the particles of proton mass carrying a negative electric charge, i.e., *negative protons*. Since a proton is 1,840 times heavier than an electron, its formation would require a correspondingly higher input of energy. It was expected that a pair of negative and positive protons should be formed when matter is bombarded by atomic projectiles carrying not less than 4.4 billion electron-volts of energy. With this task in mind, the Radiation Laboratory of the University of California in Berkeley and the construction of the gigantic electron accelerators—*Bevatron* on the West Coast and *Cosmatron* on the East Coast—that were supposed to speed up atomic projectiles to the energies necessary for the proton-pair production. The race was won by the West Coast physicists who announced in October, 1955, that they had observed negative protons being ejected from targets bombarded by 6.2 Bev (billion electron-volt) atomic projectiles. As is usual in this kind of complicated experimental research, the work was done by a team, in this case of four people: O. Chamberlain, E. Segré, C. Wiegand, and T. Ypsilantis.

The main difficulty in observing the negative protons formed in the bombarded target was that these protons were expected to be accompanied by tens of thousands of other particles (heavy mesons) also formed during the impact. Thus, the negative protons had to be filtered out separated from all the other accompanying particles. This was achieved by means of a complicated "labyrinth" formed by magnetic fields, narrow slits, etc., through which only the particles possessing the expected properties of anti-protons could pass. When the swarm of particles coming from the target (located in the bombarding beam of the Bevatron) was passed through this "labyrinth" only the negative protons were expected to come out through its opposite end. When the machine was set into operation, the four experimentalists were gratified to observe the fast particles coming out at a rate of about one every six minutes from its rear opening. As further tests have shown, the particles were genuine negative protons formed in the bombarded target by the high-energy Bevatron beam. Their mass was found to have a value of 1,840 electron masses, which is known to be the mass of an ordinary positive proton.

Just as the artificially produced positive electrons get annihilated in passing through ordinary matter containing a multitude of ordinary negative electrons, negative protons are expected to get annihilated by encountering positive protons in the atomic nuclei with which they collide. Since the energy involved in the process of proton—anti-proton annihilation exceeds, by a factor of almost two thousand, the energy involved in an electron–anti-electron collision, the annihilation process proceeds much more violently, resulting in a "star" formed by many ejected particles.

The proof of the existence of negative protons represents an excellent example of an experimental verification of a theoretical prediction concerning properties of matter, even though at the time of its proposal the theory may have seemed quite unbelievable. It was followed in the fall of 1956 by the discovery of *anti-neutrons*, i.e., the particles that stand in the same relation to ordinary neutrons as negative protons do to positive ones. Since in this case the electric charge is absent, the difference between neutrons and anti-neutrons can be noticed only on the basis of their mutual annihilation ability.

# CHAPTER 6

# Natural Radioactivity

## *Discovery and Early Progress*

The discovery of radioactivity in 1896, like that of many other unsuspected aspects of physics, was purely accidental. The French physicist, A. H. Becquerel (1852-1908), was interested at that time in the phenomenon of fluorescence, the ability of certain substances to transform ultraviolet radiation that falls on them into visible light. In one of the drawers of his desk, Becquerel kept a collection of various minerals that he was going to use for his studies, but, because of other pressing matters, the collection remained untouched for a considerable period of time. It happened that in the drawer there also were several unopened boxes of photographic plates, and one day Becquerel used one of the boxes in order to photograph something or other. When he developed the plates he was disappointed to find that they were badly fogged as if previously exposed to light. A check on the other boxes in the drawer showed that they were in the same poor condition, which was difficult to understand since all the boxes were sealed and the plates inside were wrapped in thick black paper. What could have been the cause of this mishap? Could it have something to do with one of the minerals in the drawer? Being of inquisitive mind, Becqerel investigated the situation and was able to trace the guilt to a piece of uranium ore labeled "Pitchblende from Bohemia." One must take into account, of course, that at that time uranium was not in vogue as it is today. In fact, only a very few people had ever heard about this comparatively rare and not very useful chemical element.

But the ability of uranium compounds to fog photographic plates through a thick cardboard box and a layer of black paper rapidly brought this obscure element to a prominent position in physics.

The existence of penetrating radiation that can pass through layers of ordinarily opaque materials, as if they were made of clear glass, was a recognized fact at the time of Becquerel's discovery. In fact, only a year earlier, in 1895, a German physicist, Wilhelm Roentgen (1845-1923), discovered, also by sheer accident, what are now known as X-rays, which could penetrate cardboard, black paper, or the human body equally well. But, whereas X-rays could be produced only by means of special high-voltage equipment shooting high-speed electrons at metallic targets, the radiation discovered by Becquerel was flowing steadily, without any external energy supply, from a piece of uranium ore resting in his desk. What could be the origin of this unusual radiation? Why was it specifically associated with the element uranium and, as was found by further studies, with two other heavy elements, thorium and actinium?

The early studies of the newly discovered phenomenon, which was called *radioactivity*, showed that the emission of this mysterious radiation was completely unaffected by physical and chemical conditions. One can put a radioactive element into a very hot flame, or dip it into liquid air, without the slightest effect on the intensity of the radiation it emits. No matter whether we have pure metallic uranium or its chemical compounds, the radiation flows out at a rate proportional to the amount of uranium in the sample. These facts led the early investigators to the conclusion that the phenomenon of radioactivity is so deeply rooted in the interior of the atom that it is completely insensitive to any physical or chemical conditions to which the atom is subjected.

Becquerel's discovery attracted the attention of the Polish-born Madame Marie Skodowska Curie (1867-1934), wife of the French physicist, Pierre Curie. She suspected that the radioactivity of uranium ore might, to a large extent, be due to some other chemical element, much more active than uranium, which might, however, be present in uranium ore in very small quantity. Being an experienced and hard-working chemist, Madame Curie decided to separate this hypothetical element from uranium ore by a painstaking method known as "chemical fractioning." Carloads upon carloads of pitchblende from Bohemia went through Madame Curie's chemical kitchen where careful processing was taking place, and only the fractions of the material emitting radiation were retained. Her work culminated in 1898 with a brilliant success: she obtained a few milligrams of a pure element that was a million times more radioactive than uranium itself. She christened the new element radium, and its number in the periodic system of elements was 88. Another radioactive element discovered by Madame Curie was even more active than radium and she named it "polonium" in honor of her native country. The study of radioactivity carried on by many investigators at the turn of the century led to the discovery of many other radioactive elements carrying such strange names as uranium $X_1$, ionium, radium emanation, etc. The puzzling process of radioactivity was interpreted by the British physicists, Soddy and Rutherford, to be the result of the spontaneous transformations of the elements near the end of the periodic system into other elements, in fulfillment of the dreams of medieval alchemists.

## Alpha, Beta, and Gamma Rays

In their early studies of radioactive substances, Becquerel and his followers found that their radiation consisted of three different components:

*Alpha rays*, composed of fast-moving helium nuclei. Since these nuclei carry a double positive charge and have a mass of four atomic units, the result of alpha transformation is to convert the element that emits the α-particle into another element which has an atomic number smaller by two and an atomic weight smaller by four.

*Beta rays*, composed of fast-moving negative electrons. Since a loss of negative charge is equivalent to a gain of positive charge, the atomic number of the resulting element increases by one. The atomic weight, however, does not change because of the negligibly small mass of the electron.

*Gamma rays*, which are associated both with α and β transformations, were shown to be short electromagnetic waves similar to X-rays, emitted by atomic nuclei in the process of α- or β-particle ejection.

It is quite easy (in principle, at least) to separate these three types of radiation when they are emitted from a small piece of material containing a mixture of radioactive elements. If we drill a small hole in a block of lead, which is a good absorber of radiations of all kinds, and place a speck of radioactive material at the bottom of the hole, a narrow, well-defined beam of radiation will be emitted from the top of the hole. If this beam is passed through a strong electric field between a pair of parallel plates, the single beam will be separated into its three components. The same results will follow if, instead of the electric field, the beam is passed through a strong magnetic field perpendicular to the direction of the emerging beam.

## Families of Radioactive Elements

Radioactivity, observed by Becquerel in uranium and its compounds, turned out to be a composite effect that was owing to the presence of a large number of radioactive elements, including radium and uranium themselves. In fact, studies by the British physicist, Soddy, and his famous collaborator, Rutherford, showed that this mixture contained over a dozen individual elements.

In the uranium family, which also includes radium, uranium plays the role of the head of the family and, being very long-lived, produces numerous children, grandchildren, great-grandchildren, etc. The genealogy of the uranium family is shown in Figure 6-1. An atom of UI, the father of the family ($_{92}U^{238}$; atomic number 92, mass number 238), emits an $\alpha$-particle and is transformed into an atom of so-called $UX_1$. Since the $\alpha$-particle ($_2HE^4$) carried away two units of charge and four units of mass, $UX_1$ has an atomic number of 90 (92 − 2) and a mass number of 234 (238 − 4). The element with atomic number 90 is thorium, and the so-called $UX_1$ is really an isotope of thorium, $_{90}TH^{234}$. The nuclear equation for this transformation is:

$$_{92}U^{238} \rightarrow {}_2HE^4 + {}_{90}TH^{234}$$

The next step is the emission of a $\beta$-particle by $UX_1$, which turns it into $UX_2$. The emission of a $\beta$-particle (an ordinary electron, $_{-1}e^0$) carried away a unit of negative charge, which is the same as adding a positive charge to the nucleus, and thus the atomic number is increased by one; the mass number is not changed by $\beta$-emission. Thus our $UX_1$ nucleus, which is really $_{90}TH^{234}$, becomes $UX_2(_{91}Pa^{234})$:

$$_{90}TH^{234} \rightarrow {}_{-1}\beta^0 + {}_{91}Pa^{234}$$

The next step is the emission of another $\beta$-particle by $UX_2$ which turns it into UII with the same atomic number as UI, but four units of mass lighter. The following $\alpha$-emission leads to ionium (atomic number 90, atomic weight 230), etc., etc. After seven $\alpha$-emissions and six $\beta$-emissions, we arrive at a polonium atom, which emits an eighth $\alpha$-particle and turns into an atom of lead (Pb) with atomic number $92 - 8 \times 2 + 6 \times 1 = 82$, and atomic weight $238 - 8 \times 4 = 206$. The nuclei of $Pb^{206}$ are stable and no further radioactive transformations take place.

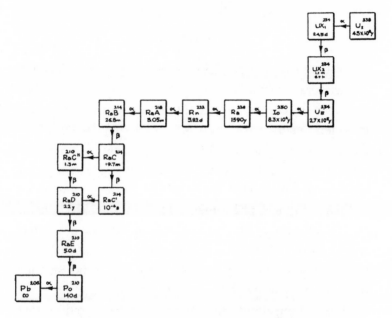

*Figure 6-1. The uranium-radium family. The numbers at the bottom of the squares give the half life periods in years (y), days (d), hours (h), minutes (m), and seconds (s). Notice the forking ($\alpha\beta$ or $\beta\alpha$) at RaC.*

Genealogically speaking, the thorium and actinium families are very similar to that of uranium and terminate with stable lead isotopes $Pb^{238}$ and $Pb^{207}$, respectively. Besides these radioactive families, which include the heaviest elements of the periodic system and are transformed by a series of intermittent $\alpha$- and $\beta$-decays into isotopes of lead, a few lighter elements also go through a one-step transformation. These include samarium ($Sm^{148}$), which emits α-rays and turns into stable $Nd^{144}$, and two β-emitters, potassium ($K^{40}$) and rubidium ($Rb^{87}$), which turn into stable isotopes of calcium ($Ca^{40}$) and strontium ($Sr^{87}$).

## Decay Energies

The velocities of α-particles emitted by various radioactive elements range from $0.98 \times 10^9$ cm/sec for samarium 148 to $2.06 \times 10^9$ cm/sec for ThC⁻; these velocities correspond to kinetic energies of from 3.2 to $14.2 \times 10^{-6}$ erg. The energies of β-particles and γ-quanta are somewhat smaller but of the same general order of magnitude. These energies are considerably higher than the energies encountered in ordinary physical phenomena. For example, the kinetic energy of atoms in thermal motion, at such a high temperature as 6,000°K (surface temperature of the sun), is only $1.25 \times 10^{-12}$ erg, i.e., several million times smaller than the energies involved in radioactive decay.

In speaking about the energies liberated in radioactive transformations, nuclear physicists customarily use a special unit known as the *electron-volt*. This unit is defined as *the energy gained by a particle carrying one elementary electric charge* (no matter whether it is an electron or any singly charged positive or negative ion) *when it is accelerated through an electric field with a potential difference of 1 volt.* Thus, the electrons accelerated in J. J. Thomson's tube, with 5,000 volts applied between the anode and cathode, acquire by this definition an energy of 5,000 electron-volts. On the other hand, the energy of a doubly charged oxygen ion, $O^{++}$, accelerated through the same potential difference will be $10^4$ electron-volts, since the electric force acting on the ions, and consequently the work done by it, is twice as large. Remembering that the value of an elementary charge on an electron, proton, or any singly charged ion is $1.60 \times 10^{-19}$ coulomb, and that a volt is one joule/coulomb, *we find that one electron-volt of energy is $1.60 \times 10^{-19}$ joule, or $1.60 \times 10^{-12}$ erg.* Another unit commonly used in nuclear work is the Mev, which stands for "million electron-volts," and the Bev, which stands for "billion electron-volts."

## Half Lifetimes

As was mentioned above, the process of natural radioactive decay is ascribed to some kind of intrinsic instability of the atomic nuclei of certain chemical elements (especially those near the end of the periodic table). From time to time this results in a violent breakup and the ejection from the nucleus of either an α-particle or an electron. The nuclei of different radioactive elements possess widely varying degrees of internal instability.

In some cases (such as uranium), radioactive atoms may remain perfectly stable for billions of years before breaking up; in other cases (such as $RaC^-$), they can hardly exist longer than a small fraction of a second. The breakup of unstable nuclei is a purely statistical process, and we can speak of the "mean lifetime" of any given element in just about the same sense as insurance companies speak of the mean life expectancy of the human population. In the case of human beings and other animals, the chance of decaying (i.e., dying) remains fairly low up to a certain age and becomes high only when the person grows old, but radioactive atoms have the same chance of breaking up no matter how long it has been since they were formed (by the decay of the previous element in the family). Since radioactive atoms begin to die out at the very moment of their birth, the decrease of their number with time is different from the corresponding decrease of the number of living individuals (Figure 6-2a and b). In the latter case, the curve of surviving individuals runs first almost horizontally and becomes steep later only when the organism begins to wear out, but the radioactive decay curve is steep all the time.

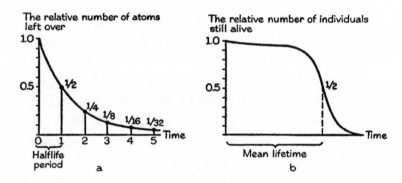

Figure 6-2. A comparison of the survival curve for radioactive atoms and for living individuals.

*The number of decaying radioactive atoms is proportional to the number of atoms available but is quite independent of the age of these atoms.*

The situation resembles that existing on a battlefield where any of the soldiers can be killed with equal chances any day of the campaign, while the cases of natural death, which depend on the soldier's age, are of very small importance. The time period during which the initial number is reduced to one-half is known as the *half-life period* of the element. At the end of twice that period, only a quarter of the original amount will be left; at the end of three half-life periods, only one-eighth will be left, etc. From the arguments above, we see a simple way to formulate mathematically the amount of a decaying element that is left after any number of half lives. If we start out with an amount of $N^0$ of some radioactive material, after n half lives have passed there will be left an amount N:

$$N = N_0 \times \left(\frac{1}{2}\right)^n$$

The gas radon, for example, has a half life of 3.8 days. If we start out with 5 mg of radon, how much will be left after a month? A month is $^{30}/3.8 = 8$ half lives, and:

$$N = 5 \times \left(\frac{1}{2}\right)^n$$
$$= 5 \times \frac{1}{256} = 0.02 \text{ mg left}$$

As we have seen above, various elements posses widely different life times (Figure 6-1). The half life of $U^{238}$ is 4.7 billion years, which accounts for its presence in nature in spite of the fact that all atoms of both stable and unstable elements may have been formed about five billion years ago, but the half life of radium is only 1,620 years; hence the 200 mg of radium separated in 1898 by Marie and Pierre Curie now contains only 195 mg. The short-lived atoms of RaC' exist, on the average, for only 0.0001 sec between the moment they are formed by β-emission of RaC and their transformation into RaD atoms.

## *Uranium-Lead Dating*

The decay of radioactive elements and its complete independence of physical and chemical conditions give us an extremely valuable method of estimating the ages of old geological formations. Suppose we pick up a rock from a shelf in a geological museum that is marked as belonging to the late Jurassic era; i.e., to the period of the earth's history when gigantic lizards were the kings of the animal world. Geologists can tell approximately how long ago this era was by studying the thicknesses of various prehistoric deposits and by comparing them with the estimated rates of the formation of sedimentary layers, but the data obtained by this method are rather inexact. A much more exact and reliable method, based on the study of the radioactive properties of igneous rocks, was proposed by Joly and Rutherford in 1913 and soon became universally accepted in historical geology. We have seen above that uranium is the father of all other radioactive elements belonging to its family and that the final product of all these disintegrations is a stable isotope of lead, $Pb^{206}$.

The igneous rock of the Jurassic era that now rests quietly on a museum shelf must have been formed as a result of some violent volcanic eruption of the past when molten material from the earth's interior was forced up through a crack in the solid crust and flowed down the volcanic slopes. The erupted molten material soon solidified into rock that did not change essentially for millions of years. But, if that piece of rock had a small amount of uranium imbedded in it, as rocks often do, the uranium would decay steadily, and the lead resulting from that decay would be deposited at the same spot. The longer the time since the solidification of the rock, the larger would be the relative amount of the deposited lead with respect to the leftover uranium. Thus, by measuring the uranium-to-lead ratio in various igneous rocks, we can obtain very exact information concerning the time of their origin and the age of the geological deposits in which they were found.

Similar studies can be carried out by using the rubidium inclusions in old rocks and measuring the ratio of leftover rubidium to the deposited strontium. This method has an advantage over the uranium-lead method because we deal here with a single transformation instead of the long sequence of transformations in the uranium family. In fact, one of the members of the uranium family is a gas (radium emanation or radon) and could partially diffuse away from its place of their formation, thus leading to an underestimation of the age of the rocks.

## Carbon Dating

Apart from the above-mentioned natural radioactive elements, which are presumably as old as the universe itself, we find on the earth a number of radioactive elements that are being continuously produced in the terrestrial atmosphere by cosmic ray bombardment. Among these, the most interesting is the heavy isotope of carbon, $C^{14}$, which is produced from atmospheric nitrogen by a high-energy neutron impact ($N^{14}$ + neutron $\rightarrow C^{14}$ + proton) and incorporated into the molecules of atmospheric carbon dioxide. Since plants use atmospheric carbon dioxide for their growth, radioactive carbon is incorporated into each plant's body, making all plants slightly radioactive throughout their life.

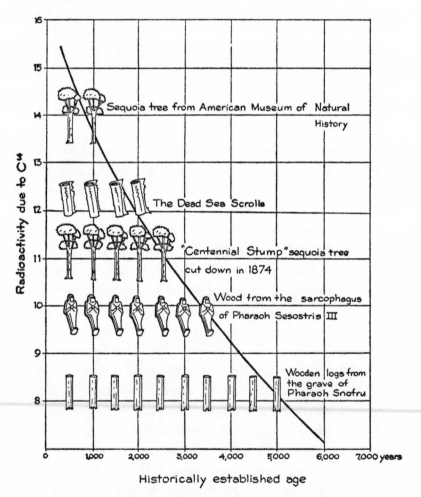

*Figure 6-3. The radioactivity, due to $C^{14}$, of various old objects as the function of their age, measured by Dr. W. Libby. Each symbol represents 500 years of age.*

As soon as a tree is cut or falls down and all of the metabolic processes stop in its body, however, no new supply of $C^{14}$ is available and the amount of radioactive carbon in the wood gradually decreases as time goes on. Since the half life of $C^{14}$ is 5,700 years, the decay will last for many millenniums, and by measuring the ratio of $C^{14}$ to the ordinary $C^{12}$ in old samples of wood we are able to estimate rather exactly the dates of origin. The studies in this direction were originated by an American physicist, W. Libby (1908–1980), and are paying the same role in the exact dating of ancient human history as the measurement of the uranium-lead ratio in the dating of the history of our globe. The measurement of $C^{14}$ radioactivity in old samples of wood is a very delicate matter since it is usually much weaker than the radioactivity of the background surrounding the object (the experimenter himself has a higher $C^{14}$ concentration than the piece of wood he is studying) and cosmic rays. Thus, the sample under investigation must be heavily shielded, and a very sensitive counter must be used. In Figure 6-3 we give a few examples of the measured and expected concentrations of radioactive carbon in various wooden objects of known age. Using these data and measuring the $C^{14}$ concentration in wooden objects of unknown age, we can easily estimate their ages. Some examples of such estimates are given in Table 6-1.

TABLE 6-1

| | |
|---|---|
| 1. Lake mud from Knock nac ran, Ireland | Age: 11,310±720 |
| 2. Wood from the bottom deposits of Lake Kickapoo, Illinois | Age: 13,842±780 |
| 3. Charcoal from Lascaux Cave near Les Eyzies, France. | Age: 15,516±900 |
| 4. Wood from the woody layer at the bottom of the sand and gravel deposit at Dyer, Indiana. | Age: 18,500±500 |

Measurements of the $C^{14}$ content of trees felled by the glaciers have established that the last glaciation of northern United States was much more recent (about 10,000 years ago) than had been previously supposed.

## Tritium Dating

Another interesting method of dating by the use of radioactive materials, which was also worked out by W. Libby, utilized the radioactivity of tritium, i.e., the heavy isotope of hydrogen with atomic weight 3. Tritium is also produced in the terrestrial atmosphere under the action of cosmic radiation and is precipitated to the surface by rains. However, tritium's half life is only 12.5 years, so that all age measurements involving this isotope can be carried out only for comparatively recent dates. The most interesting application of the tritium dating method may be in the study of the movements of water masses, both in ocean currents and in underground waters, since by taking samples of water from different locations and from different depths, we can tell by their tritium content how long ago this water came down in the form of rain.

Samples of old water are more difficult to collect than samples of old wood, and Libby resolved this problem by analyzing the tritium content in wine of different vintages, originating in different countries. Regrettably, an entire case of a fine wine must be used for each measurement and is rendered undrinkable in the process. But the agreement with the expected tritium content was in all cases excellent, as demonstrated in Figure 6-4.

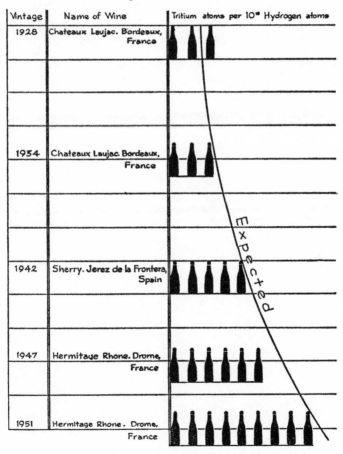

*Figure 6-4. The agreement between the calculated and the expected relationship of radioactivity vs. age in various wines.*

# CHAPTER 7

# Artificial Nuclear Transformations

## *Splitting of Atomic Nuclei*

After Rutherford became completely persuaded that the radioactive decay of heavy elements is due to the intrinsic instability of their atomic nuclei, his thoughts turned to the possibility of producing the artificial decay of lighter and normally stable nuclei by subjecting them to strong external forces. True enough, it was well known at that time that the rates of radioactive decay are not influenced at all by high temperatures or by chemical interactions, but this could be simply because the energies involved in thermal and chemical phenomena are much too small as compared with the energies involved in the nuclear disintegration phenomena. Whereas the kinetic energy of thermal motion (at a few thousand degrees) as well as the chemical energy of molecular binding are of the order of magnitude of only $10^{-12}$ erg, the energies involved in radioactive decay are of the order of $10^{-6}$ erg, i.e., a million times higher. Thus, in order to have any hope of a positive outcome, the light stable nuclei must be subjected to a much stronger external agent than just a high temperature or a chemical force, and the bombardment of light nuclei by high-energy particles ejected from the unstable heavy nuclei was the natural solution of the problem.

Following this line of reasoning, Rutherford directed a beam of α-particles emanating from a small piece of radium against a thin layer of nitrogen gas and observed, to his complete satisfaction, that besides the α-particles that passed the layer and were partially scattered in all directions, there were also a few high-energy protons (i.e., the nuclei of hydrogen) that were presumably produced in the collisions between the onrushing α-projectiles and the nuclei of nitrogen atoms. This conclusion was later supported by cloud chamber photographs, as we will discuss in the next section. The capture of an α-particle followed by the ejection of a proton increases the atomic number of the nucleus in question by one unit ($+2 - 1 = +1$) and its mass by three units ($+4 - 1 = +3$), transforming the original nitrogen atom $_7N^{14}$ into an atom $_8O^{17}$ of a heavier isotope of oxygen. We can express this reaction by the nuclear formula:

$$_7N^{14} + {}_2He^4 \rightarrow {}_8O^{17} + {}_1H^1$$

Following this original success, Rutherford was able to produce the artificial transformation of other light elements, such as aluminum, but the yield of protons produced by α-bombardment rapidly decreased with increasing atomic number of the target material, owing to the increase in electrostatic repulsion of the α-particle by the greater + charge of the larger nuclei, and he was not able to observe any ejected protons for elements heavier than argon (atomic number 18).

## Photographing Nuclear Transformations

The study of nuclear transformations was facilitated by the ingenious invention of still another Cavendish physicist, C. T. R. Wilson. This device, known as the "Wilson Chamber" or "cloud chamber," permits us to obtain a snapshot showing the tracks of individual nuclear projectiles heading for their targets and also the tracks of various fragments formed in the collision. It is based on the fact that whenever an electrically charged fast-moving particle passes through the air (or any other gas), it produces ionization along its track. If the air through which these particles pass is saturated with water vapor, the ions serve as the centers of condensation for tiny water droplets, and we see long thin tracks of fog stretching along the particles' trajectories. The scheme of a cloud chamber is shown in Figure 7-1. It consists of a metal cylinder, C, with a transparent glass top, G, and a piston, P, the upper surface of which is painted black. The air between the piston and the glass top is initially saturated with water (or alcohol) vapor, generally by a coating of moisture on top of the piston. The chamber is brightly illuminated by a light source, S, through a side window, W. Suppose now that we have a small amount of radioactive material on the end of a needle, N, which is placed near the thin window, O.

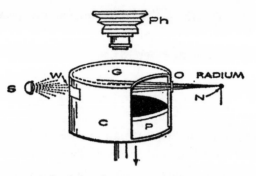

*Figure 7-1. A cloud chamber.*

The particles that are ejected by the radioactive atoms will fly through the chamber ionizing the air along their paths. However, the positive and negative ions produced by the passing particles recombine rapidly into neutral molecules. Suppose, however, that the piston is pulled rapidly down for a short distance. The rapid expansion of the air will cause it to cool, and the already saturated air now becomes supersaturated with moisture which will condense into water droplets. In order to condense, however, the droplets need centers of some sort around which to form. The natural condensation of raindrops takes place on dust particles, tiny salt crystals, or ice crystals. In *cloud-seeding*, airplanes scatter minute crystals of silver iodide to encourage the condensation of rain droplets.

In the cloud chamber, however, there is no dust, and the droplets condense on the ions (as the next best thing) that have been formed along the path of the speeding particles. Thus the tracks of the particles that passed by just before, or just as, the piston was pulled down will show up as trails of microscopic water droplets. The tracks of α-particles are quite heavy, since the massive doubly charged α ionizes the air strongly. The track of a proton is less strongly marked, and along the path of an electron the ions and hence the droplets are much more sparse. In much of the present cloud chamber work, an intense magnetic field is created within the chamber, so the charged particles are deflected into curved paths. By measuring the curvature shown on the photographs, we can compute the speed and energy of the particles.

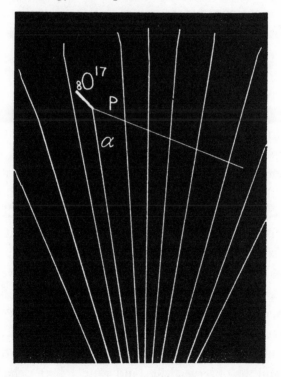

*Figure 7-2. Sketch of the first cloud chamber photograph, taken by Blackett, of nuclear disintegration. The long thin track at upper right is that of a proton that has been ejected from a nitrogen hit by an α-particle. The short thicker track, at upper left, is that of an oxygen nucleus formed in the collision. All other α-particles shown in the sketch lost their energy before they had a chance to hit a nucleus.*

Figure 7-2 is a sketch of the classical cloud chamber photograph, taken in 1925 by P. T. M. Blackett, that shows the collision of an incident $\alpha$-particle with the nucleus of a nitrogen atom in the air which fills the chamber. The long thin track going almost backward is that of a proton ejected in that collision. It can be easily recognized as a proton track because protons are four times lighter than $\alpha$-particles and carry only one-half as much electric charge; therefore they produce fewer ions per unit length of their path than $\alpha$-particles. The short heavy track belongs to the nucleus $_8O^{17}$ formed in the process of collision.

## Bubble Chambers

In recent years, the "bubble chamber" has been developed to supplement the work of the cloud chamber. Although the general principle of its operation is the same as that of the cloud chamber, the bubble chamber is filled with a liquid (often propane and, more recently, liquid hydrogen) which is kept exactly at its boiling-point temperature. A slight expansion will reduce the pressure on the liquid, and bubbles of vapor will form on the ions which have been produced in the liquid by passing particles. The bubble chamber has a great advantage when the collision events to be observed are relatively rare. In the closely packed atoms of a liquid, many more nuclear collisions will occur than in a gas, and the observer will stand a much better chance of photographing what he is looking for than he would with a cloud chamber.

## First Atom Smashers

Since the only massive projectiles emitted by the nuclei of natural radioactive elements are α-particles, i.e, the nuclei of helium, it was desirable to develop a method for the artificial production of beams formed by other atomic projectiles, particularly beams of high-energy protons. According to theoretical considerations, the ease with which a bombarding particle penetrates into the structure of a bombarded atomic nucleus depends on the atomic number (i.e., the nuclear electric charge) of the element in question. The larger the atomic number, the stronger is the electric repulsive force opposing the approach of α-particles to the nucleus, and, consequently, the smaller are the chances of a demolishing collision. Since protons carry only one-half of the electric charge carried by an α-particle, they were expected to be much better as atomic projectiles and to be able to smash atomic nuclei of light elements even when moving with only 1 Mev of energy. Rutherford asked John Cockcroft to construct a high-tension machine that would accelerate protons to the energy of 1 million electron-volts, and, within a couple of years, the first "atom smasher" was constructed by Cockcroft and his associate, E. T. S. Walton. Directing the beam of 1-Mev protons at a lithium target, Cockcroft and Walton observed the first nuclear transformation caused by artificially accelerated projectiles. The equation of this reaction is:

$$_3Li^7 + {}_1H^1 \rightarrow 2\,_2He^4$$

If we use boron instead of lithium as the target, the reaction will be:

$$_5B^{11} + {}_1H^1 \rightarrow 3\,_2He^4$$

## *The "Van de Graaff"*

Cockcroft and Walton's atom smasher, which was based on the electric transformer principle, gave rise to a series of ingenious devices for producing high-tension beams of atomic projectiles. The *electrostatic atom smasher* constructed by R. Van de Graaff (1901–1967) and usually called by his name, is based on a classical principle of electrostatics, according to which an electric charge communicated to a spherical conductor is distributed entirely on its surface. Thus, if we take a hollow spherical conductor with a small hole in its surface, insert through this hole a small charged conductor attached to a glass stick, and touch the inside surface of the sphere (Figure 7-3a), the charge will spread out to the surface of the big sphere.

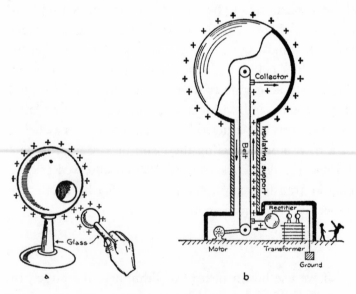

*Figure 7-3. The principle (a) and the actual scheme (b) of Van de Graaff's high-tension machine.*

Repeating the operation many times, we will be able to communicate to the large conductor any desirable amount of electricity and raise its electric potential as high as desired (or, at least, until the sparks start jumping between the conductor and the surrounding walls).

In the Van de Graaff atom smasher (Figure 7-3*b*), the small charged ball is replaced by a continuously running belt that collects electric charges from a source at the base and deposits them on the interior surface of the large metallic sphere. The high electric potential developed in this process is applied to one end of an accelerating tube in which the ions of different elements are speeded up to energies of many millions of electron-volts.

## *The Cyclotron*

Another popular atom smasher, invented by E. O. Lawrence (1901–1958), is based on an entirely different principle and utilized the multiple acceleration of charged particles moving along a circle in a magnetic field. The principle of the cyclotron is shown in Figure 7-4. It consists essentially of a circular metal chamber cut into halves, $C_1$ and $C_2$, and placed between the poles of a very strong electromagnet. The half-chambers, $C_1$ and $C_2$, are connected with a source of alternating high potential, *AC*, so that the electric field along the slit separating them periodically changes its direction. The ions of the element to be used as atomic projectiles are injected in the center of the box, *I*, at a comparatively low velocity, and their trajectories are bent into small circles by the field of the magnet. The gimmick of the cyclotron is that, for a given magnetic field, the period of revolution of an electrically charged particle along its circular trajectory is independent of the velocity with which that particle is moving. Since the increase in the radius of the path and the length of the circular trajectory is exactly proportional to the increase in velocity, the time necessary for one revolution remains the same.

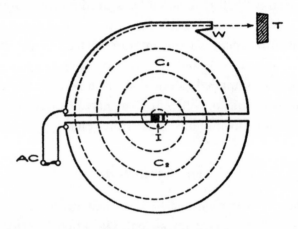

*Figure 7-4. The principle of the cyclotron.*

If things are arranged in such a way that the period of revolution of the ions injected into the field of the magnet is equal to the period of alternating tension produced by the *AC* source, the particles arriving at the boundary between the two half-chambers, $C_1$ and $C_2$, will be subject each time to an electric force acting in the same direction that the particles are moving. Thus, each time the ion passes through that boundary, it will be given additional acceleration and its velocity will gradually increase. Gathering speed, the ions will move along an unwinding spiral trajectory and will finally be ejected through the window, *W*, in the direction of the target, *T*. The largest existing cyclotron is in the radiation laboratory of the University of California. It has an accelerating circular box 60 inches in diameter and produces artificial α-beams with an energy of 40 Mev (4.5 times higher than that of the fastest natural α-particles). With this atom smasher, it was possible to cause the artificial transformation of all elements up to the heaviest ones.

# CHAPTER 8

# The Structure of the Atomic Nucleus

## *Nuclear Constituent Particles*

As was mentioned earlier, the fact that the atomic weights of isotopes of different elements are very closely represented by integral numbers strongly suggests that atomic nuclei are formed by adding together the nuclei of hydrogen atoms, i.e., protons. But if the atomic nuclei were formed exclusively by proton aggregates, they would have a much larger electric charge than that actually observed. For example, the nucleus of the oxygen atom has a mass very close to that of 16 protons, but carries only 8 elementary units of positive charge. It follows that 8 of the 16 hydrogen nuclei that combine to form the nucleus of oxygen have lost their positive charge, i.e., have turned into neutrons. Thus, the composition of the oxygen nucleus can be written as:

$$_8O^{16} = 8 \text{ protons} + 8 \text{ neutrons}$$

Similarly, we may write for the composition of the nucleus of the principal isotope of iron:

$$_{26}Fe^{56} = 26 \text{ protons} + 30 \text{ neutrons}$$

and, for the principal isotope of uranium:

$$_{92}U^{238} = 92 \text{ protons} + 146 \text{ neutrons}$$

The assumption that atomic nuclei are aggregates of protons and neutrons is substantiated by the fact that both protons and neutrons are observed to be ejected from atomic nuclei in the processes of artificial nuclear transformations. According to this assumption, α-particles emitted by various radioactive elements are composite units formed by two protons and two neutrons each. It is believable that α-particles do not exist inside atomic nuclei as such, but are built up from nuclear protons and neutrons just before their emission. Conversely, an α-particle entering the nucleus breaks up into two protons and two neutrons, which then mix with the other nuclear constituent particles.

We notice that, whereas in the case of light nuclei, the number of neutrons is equal to the number of protons, the number of neutrons exceeds the number of protons for heavier elements; the number of neutrons is 20 percent larger in iron and 50 percent larger in uranium. Neutrons outnumber protons in the heavier nuclei because the protons, being positively charged, repel each other, and their relative number must be reduced in order to secure the stability of the nucleus.

## *The Nucleus as a Fluid Droplet*

It is clear that the forces which hold the nucleus in one piece cannot be of a purely electric nature, since half of the nuclear particles (neutrons) do not carry any electric charge, whereas the other half (protons) are all positively charged, thus repelling one another and contributing to nuclear disruption rather than to stability. In order to understand why the constituent parts of the nucleus stick closely together, we must assume that there exist between them forces of some kind, attractive in nature, which act on uncharged neutrons as well as on positively charged protons. These forces that make them adhere, irrespective of the nature of the particles involved, are generally known as *cohesive forces* and are encountered, for example, in liquids, where they hold together the separate molecules and lead to the familiar phenomenon of surface tension.

A particle deep inside a liquid is subjected to attractive forces from the other particles surrounding it on all sides, so that the total resultant of all these pulls is exactly zero. On the other hand, a particle located on the surface has all its neighbors on only one side, so that their combined attraction results in a strong force pulling the particle inward. Since every particle located on the surface is pulled inward by other particles, the liquid will have a general tendency to reduce its free surface to the smallest possible value, which explains the spherical shape assumed by any liquid drop that is not acted upon by any external forces; it is well known that the sphere is the geometrical figure that possesses the smallest surface for a given total volume.

The assumption that the forces acting between the constituent  particles of the nucleus are similar to those acting between the molecules of any ordinary liquid leads to the *droplet model* of an atomic nucleus, according to which different nuclei are considered as minute droplets of a universal *nuclear fluid*.

The first important consequence of the nuclear-droplet theory is the conclusion that the volumes of different atomic nuclei must be proportional to their masses, since the density of the fluid always remains the same, regardless of the size of the droplet which it forms. This conclusion is completely confirmed by direct measurements of nuclear radii of atomic nuclei vary as the cube roots of their masses. Thus, the radii of the atomic nuclei of oxygen and lead, which have masses of 16 and 206 atomic mass units, respectively, are measured to be $3 \times 10^{-13}$ and $7 \times 10^{-13}$ cm. From these figures we see that the lead nucleus is 13 times as massive as the oxygen nucleus; hence it has volume 13 times as great. The cube root of 13 is 2.35, which is exactly the ratio of the nuclear radii. Remembering that the actual mass of the hydrogen atom is $1.66 \times 10^{-24}$ gm, we find that the nuclei of oxygen and lead have masses of $2.66 \times 10^{-23}$ gm and $3.42 \times 10^{-22}$ gm. Since the volume of these two nuclei are $1.13 \times 10^{-37}$ and $1.44 \times 10^{-36}$ cm$^3$, we obtain in both cases a value of nearly $2.4 \times 10^{14}$ gm/cm$^3$ for the density of nuclear fluid. This is a very high density indeed! If the nuclear fluid, which is dispersed through space in the form of minute droplets surrounded by rarefied electronic envelopes, could be collected to form a continuous material, 1 cc of it would weigh 240 million tons.

Along with its almost unbelievably high density, the nuclear fluid possesses a correspondingly high surface tension. The surface tension of a liquid is characterized by the force acting on a unit length of the free surface boundary. For example, if we spread a soap film over the area enclosed by a U-shaped wire, with a piece of straight wire put across it, the forces of surface tension will pull the cross wire in an attempt to reduce the surface of the film. Measuring this force and dividing it by the length of the cross wire, and also by the factor 2 because the soap film has two surfaces and consequently acts with a double force, we arrive at the value of the surface tension force acting on the unit length. The surface tension of water, for example, is known to be about 75 dynes/cm at room temperature. The surface tension of nuclear fluid is found to be 93,000,000,000,000, 000,000 dynes/cm, which is as nice and big a number as the one which describes the density of the fantastic nuclear fluid. A nuclear film attached to a wire 1 cm long would support the weight of 1 billion tons.

The value of the surface tension force immediately determines the amount of energy that is connected with any change of the total free surface of a fluid. In fact, any increase of the free surface requires work to be done against the surface tension forces, a decrease of the surface area will, on the contrary, liberate a certain amount of energy. Numerically, the amount of energy in ergs per $cm^2$ of surface area is given by the same number as the surface tension force, in dynes/cm, and is equal to $9.3 \times 10^{19}$ ergs/$cm^2$, so that, in order to calculate the total surface energy of the nucleus, we have to multiply the surface area by the above number.

Instead of expressing the surface energy per unit of surface area, we can more conveniently express it per single nuclear particle located on the surface. Since the diameter of a neutron or proton is about $3.2 \times 10^{-13}$ cm, each of them occupies on the nuclear surface an area of about $10^{-25}$ cm$^2$, so that there are $10^{25}$ particles per cm$^2$. Dividing the above total surface energy per cm$^2$ by the number of the particles per cm$^2$, we find that there is about $9 \times 10^{-6}$ erg, or about 5 million electron-volts, of energy per particle. This represents the energy that would be necessary to remove one nucleon from the surface of the nucleus against the forces of cohesion, and is analogous to the heat of evaporation per molecule for ordinary liquids.

## Mass Defect and Nuclear Binding Energy

In comparing the masses of various atomic nuclei with the masses of the protons and neutrons from which they are formed, we always find a slight discrepancy. For example, in the case of oxygen, we have:

$$
\begin{aligned}
8 \text{ neutrons} &= 8 \times 1.00898 = & 8.07184 \\
8 \text{ protons} &= 8 \times 1.00759 = & 8.06072 \\
8 \text{ electrons} &= 8 \times 0.00055 = & \underline{0.00440} \\
& & 16.13696
\end{aligned}
$$

which may be compared with the atomic weight of oxygen, or 16.00000.*

*This figure is an exact integer not because of any property of the oxygen atom but because atomic weight figures are based on a scale in which the atomic weight of the principal isotope of oxygen is assumed to be exactly 16.

It was necessary to add the 8 electrons because atomic weights always include the weight of the atomic electrons in their values. Since the number of atomic electrons is always equal to the number of protons in the nucleus, it may be more convenient, in figuring nuclear discrepancies, to use the hydrogen atom instead of the proton, as this automatically includes the proper number of electron masses. For oxygen, we could have written:

$$
\begin{array}{lll}
8 \text{ neutrons} & = 8 \times 1.00898 = & 8.07184 \\
8 \text{ hydrogens} & = 8 \times 1.00814 = & \underline{8.06512} \\
& & 16.13696
\end{array}
$$

Thus, by either method, the oxygen nucleus is seen to be lighter than its constituents by 0.13696 units of atomic weight, or atomic mass units (amu).

Similarly, in the case of the principal isotope of iron, we have:

$$
\begin{array}{lll}
8 \text{ neutrons} & = 30 \times 1.00898 = & 30.26940 \\
8 \text{ hydrogens} & = 26 \times 1.00814 = & \underline{26.21164} \\
& & 56.48104 \text{ amu}
\end{array}
$$

which is to be compared with the exact value of 55.9571 for the atomic weight of $Fe^{56}$. Here we find the atomic weight of $Fe^{56}$ is 0.5239 amu smaller than the combined masses of its components.

The explanation of this so-called mass defect* is based on Einstein's law of the equivalence of mass and energy.

---

*In Aston's original definition, *mass defect is the difference between the exact value of the atomic weight and the nearest integral number*. However, for theoretical considerations, it is more rational to use the definition of mass defect as given here.

When a nucleus is being formed from individual protons and neutrons, large amounts of nuclear energy are set free, just as chemical energy is liberated in the form of heat during the formation of water molecules from hydrogen and oxygen atoms. This energy possesses a certain mass, which is carried away and makes the resultant nucleus correspondingly lighter. Since the unit of atomic weight ($1/16$ of the actual weight of the oxygen atom) equals $1.66 \times 10^{-24}$ gm, its energy equivalent is:

$$1.66 \times 10^{-24} (3 \times 10^{10})^2 = 1.48 \times 10^{-3} \text{ erg or } 932 \text{ Mev}$$

Dividing the total binding energy of the composite nucleus by the total number of protons and neutrons forming it, we obtain a mean binding energy per particle.*

In Figure 8-1, we give a plot of the nuclear binding energy per particle against atomic weight. We notice that this binding energy per particle (which has rather large values for light nuclei) decreases with the atomic weight, reaches a minimum in the neighborhood of A=50, and then begins to increase again toward the heavy elements.

The reason for this behavior of binding energy is the fact that there are two opposing forces acting within the nucleus:

1. The nuclear attractive forces that attempt to hold the nuclear constituent particles together.

2. Electrostatic repulsive forces between protons that push them apart.

---

*If we use Aston's definition of mass defect, the mass defect per nuclear particle is known as the *packing fraction*.

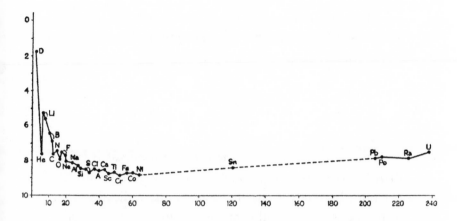

Figure 8-1. *The binding energy per nucleon vs. atomic mass number.*

In the case of lighter elements, nuclear attraction between the constituent particles prevails over the electrostatic repulsion between the protons, and the increase of the total number of particles leads to the strengthening of their mutual binding. In the case of heavier elements, the effect of repulsion between protons becomes more noticeable, weakening the average binding between the nuclear particles.

## Mass Defect and Nuclear Reactions

The exact knowledge of the atomic weights of the isotopes permits us to evaluate the energy balance of various nuclear reactions, since the mass equivalent of the liberated or absorbed nuclear energy must enter into the equation of the conservation of mass during the transformation. Thus, in the case of Rutherford's original reaction:

$$_7N^{14} + {}_2He^4 \rightarrow {}_8O^{17} + {}_1H^1$$

The sums of masses of the atoms entering the reaction and those resulting from it are, respectively:

$$_7N^{14} \longrightarrow 14.00753 \qquad _8O^{17} \longrightarrow 17.00450$$
$$+ \qquad\qquad\qquad +$$
$$_2He^4 \longrightarrow \underline{4.00386} \qquad _1H^1 \longrightarrow \underline{1.00813}$$
$$\phantom{_2He^4 \longrightarrow} 18.01139 \qquad\qquad\qquad 18.01263$$

The combined mass of the reaction products is larger than the combined mass of the atoms entering into the reaction, by 0.00124 atomic mass units, indicating that some energy is lost in this process. Using the mass-energy conversion factor given in the previous section, we obtain for the energy balance:

$$\Delta E = -1.84 \times 10^{-6} \text{ erg} = -1.15 \text{ Mev}$$

which coincides with the difference between the kinetic energy of the incident $\alpha$-particle and the kinetic energy of the ejected proton as observed by Rutherford. On the other hand, Cockcroft-Walton's reaction:

$$_3Li^7 + _1H^1 \rightarrow 2_2He^4$$

leads to the following:

$$_3Li^7 \longrightarrow 7.0182 \qquad _2He^4 \longrightarrow 4.0039$$
$$+ \qquad\qquad\qquad +$$
$$_1H^1 \longrightarrow \underline{1.0081} \qquad _2He^4 \longrightarrow \underline{4.0039}$$
$$\phantom{_1H^1 \longrightarrow} 8.0263 \qquad\qquad\qquad 8.0078$$

In this case, the difference is +0.0191 atomic mass units, corresponding to an energy liberation of $1.8 \times 10^{-5}$ erg or 17.7 Mev per reaction.

## *Nuclear Shell Model*

As we explained in the earlier discussion of the Bohr atom (under Electron Shells and the Periodic System), the regular repetition of the chemical properties of atoms arranged in order of increasing atomic number is due to the formation of the consecutive electron shells in the electronic envelopes of the atoms. Similar periodic changes are also observed in the case of atomic nuclei, manifesting themselves in the behavior of nuclear binding energies, magnetic properties, ability to participate in various nuclear reactions, etc. Inspecting the binding energy curve, we notice that, instead of being smooth, it contains kinks (at O, Cr, and other points), indicating that certain irregularities in the internal nuclear structure are present. Another example of these irregularities is shown in Figure 8-2, where the ability of different nuclei to capture neutrons (capture cross sections) is plotted against atomic weight. We notice that the regular increase of this ability with increasing atomic weight is interrupted by sharp minima (at He, O, and Cr).

Detailed studies of these and other irregularities of nuclear properties led to the conclusion that they always occur when either the number of neutrons or the number of protons is one of the following numbers: 2, 8, 14, 20, 28, 50, 82, 126, which represent the number of particles at which nuclear shells are completed. These so-called magic numbers are analogous to the sequence of numbers: 2, 10, 18, 36, 54, etc. (atomic numbers of the rare gases), that characterize the periodic system of chemical elements and that represent the number of electrons at which atomic shells are completed. The abnormally small neutron-capture cross sections for the elements with completed neutron shells (Figure 8-2) are analogous to the chemical inertness of the rare gas atoms that possess completed electron shells.

There are, however, two important differences between the shell structure of nuclei and the shell structure of atoms. In atoms, one system of shells accommodates the electrons of the atomic envelope, but in nuclei there are two independent sets of shells: one for neutrons and one for protons. Another difference lies in the fact that, whereas the electron shells in the atom are geometrically separated, nuclear shells apparently interpenetrate each other and so can be distinguished only by their different energies.

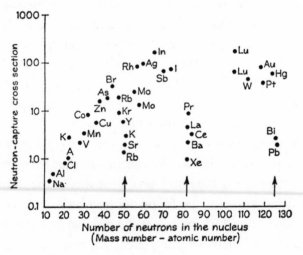

Figure 8-2. A cross-section graph of neutron capture.

## Fusion and Fission

In considering atomic nuclei of different chemical elements as minute droplets of universal nuclear fluid, we may expect that these nuclear droplets will behave in about the same way as droplets of ordinary liquid. In observing droplets of, let us say, mercury rolling on the surface of a saucer, we notice that whenever two droplets meet, they fuse together, forming a larger droplet. The fusion of two droplets into one is the work of surface tension forces, which tend to reduce the total free surface of the liquid. In fact, it is easy to show that the surface tension forces, which tend to reduce the total free surface of one big droplet is smaller than the combined surfaces of two half-size droplets. Since the total volume of the compound droplet is twice the volume of each of the smaller ones, its radius must be $\sqrt[3]{2}$ = 1.26 times larger, and its surface $(1.26)^2$ = 1.59 times larger. Thus, when two half-size droplets fuse into one, the total surface reduces in the ratio 2:1.59, or by 20 percent. It is not difficult to show by simple arithmetic that the same is true when the two droplets are of different size, although in this case the relative decrease of the total surface will be smaller. The fusion of two droplets with a mass ratio 3:1 reduces the surface by 19 percent, whereas the mass ratio 10:1 leads to a surface reduction of only 13 percent. Thus, the fusion of two droplets into one always leads to the liberation of surface energy and always takes place spontaneously whenever two droplets come into contact. If the surface tension forces were the only forces acting in atomic nuclei, any two nuclei would fuse together, liberating nuclear energy.

However, the situation changes quite considerably if we take into account that, apart from surface tension forces, electric forces of repulsion are also present in the nuclei. In contrast to ordinary liquids, a nuclear fluid is always electrically charged, since about one-half of its constituent particles are protons. The electric repulsion between the nuclear charges acts in the opposite direction to the surface tension forces and tends to disrupt larger droplets into smaller ones. In order to calculate the change of electric energy connected with the fission of a nucleus into two halves, we must make use of the fact that the electrical energy of a charged sphere is proportional to the square of its charge divided by its radius. Since each of the two resultant droplets has one-half of the original charge and their radii are 1.26 times smaller than that of the original droplet, the electric energy of each will be $(^1/_2)^2 \times 1.26 = 0.315$ times that of the original big droplet. The combined electric energy of two halves will be only 0.63 of the original value. Thus we conclude that electrostatic forces act in the opposite direction to the surface tension forces, favoring nuclear fission rather than fusion.

With both kinds of forces present, the answer to the question whether nuclear energy will be liberated in fusion or fission depends on the relative strength of the two forces. For the nuclei in which surface tension forces are stronger than electrostatic forces, fusion is an energy-liberating process. If, however, the reduction of electric energy over-balances the increase of surface energy, the fission process is to be expected. If we proceed along the sequence of elements from the lighter nuclei to the heavier ones, the surface energy, which is determined by the total surface of the nucleus, increases comparatively slowly, being proportional to the two-thirds power of the atomic weight (because of the constancy of nuclear density). On the other hand, electric energy increases approximately as the square of the nuclear charge or, what is about the same, as the square of the atomic weight.

For light nuclei, the surface tension energies (which give a release of energy by fusion) overshadow the effects of electric charge, *so the fusion of two light nuclei will liberate excess energy* as a by-product. However, since electric energy increases with atomic weight much faster than the surface energy does, we should expect the situation to be reversed for heavy nuclei, so that the electric charge factor is of greater importance. For these nuclei of high atomic number, we would thus expect the energy that is released by splitting the electric charge of the nucleus in two to be greater than the energy used up because of the greater surface area of the two fragments. This reasoning would lead us to predict that *excess energy will be released when a large nucleus fissions, or splits in two.*

This theoretical conclusion is in complete agreement with the empirical evidence given by the study of nuclear binding energies. Looking at Figure 8-1 we notice that the value of the binding energy per particle decreases with the atomic weight for the elements in the earlier (lighter) part of the periodic system. This means that if two nuclei belonging to this region fuse together, a certain amount of nuclear energy will be set free. On the other hand, for the later (heavier) part of the periodic system, the binding energy per particle increases with increasing atomic weight, indicating that for these elements fission and not fusion will be the energy-liberating process. In between these two regions lie the elements in the neighborhood of the iron group, which have the maximum binding energy per particle and are therefore stable with respect to both fusion and fission.

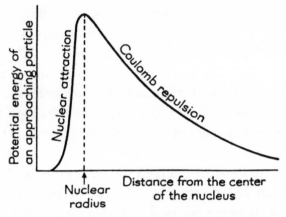

Figure 8-3. The potential energy of a positively charged particle (α-particle, proton, etc.) in the neighborhood of an atomic nucleus.

## Nuclear Potential Barrier

When a positively charged nuclear projectile, such as an α-particle or a proton, approaches an atomic nucleus, it is acted upon by electrostatic repulsive forces and cannot come into direct contact with the nucleus unless its kinetic energy is large enough to overcome the repulsion. However, as soon as the contact is achieved, nuclear attractive forces take hold of the approaching particle and pull it into the nucleus. Plotting the potential energy of a positively charged particle in the neighborhood of the atomic nucleus, we obtain the curve shown in Figure 8-3. This curve represents a "potential barrier" for the penetration of the incident positively charged particles into the nucleus, as well as for the escape of such particles from the nucleus. According to classical mechanics, the incoming and outgoing nuclear particles can pass the potential barrier only if their kinetic energy is larger than the maximum height of the barrier. Experimental evidence shows, however, that this is definitely not so. The uranium nucleus, for example, has a radius of $9 \times 10^{-13}$ cm and is surrounded by a potential barrier 27 Mev high. We would expect, therefore, that only particles having 27 Mev of potential energy or more would be able to escape from the uranium nucleus. We know, however, that $\alpha$-particles emitted by uranium have an energy of only 4 Mev, and it is difficult to understand how they get out across the barrier at all. Also, in the case of the artificial transformation of elements, such as Rutherford's experiments on the bombardment of nitrogen by $\alpha$-particles, the energy of the projectiles is often lower than the height of the potential barrier surrounding the bombarded nucleus; nevertheless, these projectiles penetrate into the nuclear interior, causing its disintegration.

## Tunnel Effect

This paradoxical phenomenon, known as the *tunnel effect*, was explained in 1928 by G. Gamow and, independently, by R. Gurney and E. Condon, as an outcome of the wave nature of nuclear particles. In order to understand the

situation, let us consider a simple example from the field
of optics. A light beam falling on the interface between
a dense and a light material (passing from glass into air
for example), will be refracted with an angle of refraction
larger than the angle of incidence. If, however, the angle
of incidence exceeds a certain value, the phenomenon of
"total internal reflection" will take place, and no light at
all will penetrate into the second medium. This occurs
because the equation:

$$\frac{\sin i}{\sin r} = \frac{1}{n} < 1$$

has no solution for $r$ when $i$ exceeds a certain critical value,
since no angle has a sine greater than 1. If, however, we
look at this phenomenon from the point of view of wave
optics, we find that it is considerably more complicated
than it is represented to be in geometrical optics.

Indeed, it can be shown that in the case of total internal
reflection, the incident light waves are not reflected entirely
from the geometrical surface separating glass and air, but
penetrate into the air to a depth of several wave lengths.
The line of flow of energy for this case are shown in
Figure 8-4a. We see that, on passing through the interface,
the original light beam breaks into many components
which penetrate into the air to different depths but always
come back into the glass to form the reflected beam. This
phenomenon cannot be described in terms of geometrical
optics and should be considered as a peculiar case of the
diffraction of light.

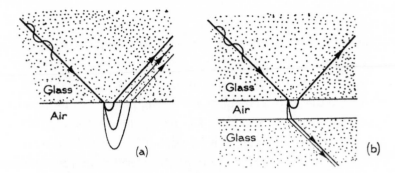

*Figure 8-4. The total internal reflection of light from the glass-air interface (a), and a partial penetration into another piece of glass placed within a few wave lengths of the first one.*

If we place another piece of glass right under the first one (Figure 8-4*b*) so that the thickness of the air layer between them will be equal to only a few wave lengths, some of the light entering the air layer will reach the surface of the second piece of glass and form in it a light beam parallel to the incident beam. The intensity of that beam decreases very rapidly with increasing thickness of the air layer and becomes negligibly small when this thickness exceeds several wave lengths. Thus, wave optics explains a phenomenon which would be completely inexplicable from the point of view of classical geometrical optics.

Let us now consider a material particle with an initial energy, $E$, which falls on a potential barrier the height of which, $U$, is larger than the energy of the particle. According to classical mechanics, this particle cannot enter the region occupied by the potential barrier, since its kinetic energy within the barrier would be:

$$^1/_2 m\upsilon^2 = E - U < 0$$

which has no real solution for $v$. The situation is different, however, if we consider the motion of the particle to be guided by de Broglie waves; the potential barrier plays the same role for the de Broglie waves as the air layer between the two pieces of glass plays for the light waves in the case of total internal reflection. The de Broglie waves incident on a potential barrier will be partially reflected from its outer boundary while a part of them will penetrate into the barrier itself (Figure 8-5). The part that penetrates into the barrier will reach its other side and will come out into the region beyond the barrier. Since the propagation of de Broglie waves guides the motion of material particles, it follows that some of the particles falling on the barrier will pass through it even though this contradicts classical mechanics. It should be noticed that the intensity of the de Broglie waves that pass through the barrier will become negligibly small if the length of the barrier exceeds several wave lengths. Since the number of particles guided by de Broglie waves is proportional to their intensity, we must conclude that the number of particles which pass through the barrier will in this case also be very small.

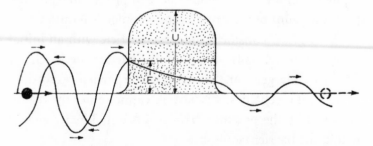

Figure 8-5. A de Broglie wave falling on a potential barrier, the height of which, U, is larger than the kinetic energy, E, of the particle. The wave is partially reflected and it partially passes through.

## *Alpha Decay and Nuclear Bombardment*

As we mentioned before, α-particles can escape from the
nuclei of radioactive elements only if they are able to pass
through the nuclear potential barriers, the height of which
exceeds many times the energy of escaping particles. Since
such a feat is absolutely impossible from the point of view
of classical mechanics, the phenomenon of radioactivity
would not exist if the laws of classical mechanics were
completely unshakable. We have seen, however, that the
leakage of the de Broglie waves through potential barriers
of any height opens the possibility for the escape of α-
particles from the nuclei, even though the chances of such
an escape may be extremely low. It has been calculated
that an α-particle making an attempt to cross the potential
barrier surrounding a uranium nucleus has only one
chance in $10^{38}$ to do so. Incredibly small as this chance
seems to be, success can be finally achieved if a sufficiently
large number of attempts is made. Let us remember that
the α-particles imprisoned in the nuclear interior are
restlessly rushing to and fro, and constantly bouncing
from the high walls of the nuclear potential barrier that
surrounds them on all sides. Each time an α-particle hits
the wall of its prison, it has a slight chance (one out of
$10^{38}$) to get out. How many such escape attempts are made
per second? The velocity with which the imprisoned α-
particles move inside their nuclear prison is of the order
of $10^9$ cm/sec, while the size of the prison is about $10^{21}$
times per second. Since the chance of escape in any single
collision is only one out of $10^{38}$, $10^{38}$ escape attempts must
be made altogether. At the rate of only $10^{21}$ attempts per

second, $\alpha$-particles must go on trying for $10^{38}/10^{21} = 10^{17}$ sec = $3 \times 10^9$ years. And indeed, as we have seen before, the half lifetime of uranium nuclei is measured in billions of years.

Let us now take the case of RaC' which, in contrast to uranium, has a half-life period of only 0.0001 sec. Why does this nucleus decay so much faster? A detailed study shows that there are two reasons for it: first, the electric charge of the RaC' nucleus is smaller than that of the uranium nucleus, which reduces the height of the potential barrier through which the escaping $\alpha$-particle must penetrate; secondly, the energy of $\alpha$-particles from RaC' is almost twice as large as that of $\alpha$-particles from uranium. Carrying out the same kind of calculations as we did in the case of the uranium nucleus, we find that an $\alpha$-particle escaping from a RaC' nucleus must make only $10^{17}$ attempts to have a good chance to get away. At the rate of $10^{21}$ attempts per second, the mean waiting time for the escape reduces to $10^{17}/10^{21} = 0.0001$ sec.

Thus, we see that comparatively small variations in the height of the potential barrier and in the velocity of escaping $\alpha$-particles can change the half lifetime periods from billions of years to small fractions of a second. This accounts for the great variability of the half-life periods among the radioactive elements.

What is true for the $\alpha$-particles escaping from radioactive nuclei is also true for $\alpha$-particles, and other positively atomic projectiles, that are shot at the nuclei of the ordinarily stable elements. In order to penetrate the nuclei and to cause some kind of nuclear reaction, these projectiles must first penetrate the high potential barrier surrounding the bombarded nucleus. When Rutherford bombarded nitrogen nuclei by α-particles, and when Cockcroft and Walton bombarded lithium nuclei by artificially accelerated protons, the energy of the projectiles was always smaller than the heights of the potential barriers surrounding the nuclei in question. Thus, the success of these experiments was entirely due to the quantum mechanical tunnel effect.

# CHAPTER 9

# Large-Scale Nuclear Reactions

## *Discovery of Fission*

Neutrons are the ideal projectiles for nuclear bombardment because they have no electrical charge and thus suffer no repulsion in their approach to atomic nuclei. Following the discovery of neutrons, many new types of artificial nuclear transformations have been investigated. In some cases, the impact of a neutron may result in the ejection of a proton or an α-particle, as in the reactions:

$$_7N^{14} + {}_0n^1 \rightarrow {}_6C^{14} + {}_1H^1$$
$$_7N^{14} + {}_0n^1 \rightarrow {}_5B^{11} + {}_2He^4$$

In some cases, the incident neutron can eject another neutron without being captured itself:

$$_6C^{12} + {}_0n^1 \rightarrow {}_6C^{11} + 2{}_0n^1$$

whereas in other cases the incident neutron can be captured by the nucleus with the release of excess energy in the form of γ-quantum. The latter process, known as the *radiative capture* of neutrons, is of particular importance for heavy nuclear targets, since in this case the ejection of protons and $\alpha$-particles is strongly hindered by the "outgoing" potential barrier surrounding the nucleus. The radiative capture of the neutron leads to the formation of a heavier isotope of the bombarded element. Sometimes these isotopes are stable, so that no further nuclear transformation takes place:

$$_8O^{16} + {_0}n^1 \rightarrow {_8}C^{11} + \gamma$$

whereas in some other cases the radiative capture of a neutron leads to a $\beta$-emission:

$$_{47}Ag^{109} + {_0}n^1 \rightarrow {_{47}}Ag^{110} + \gamma$$
$$_{47}Ag^{110} \rightarrow {_{48}}Cd^{110} + e^-$$

which is necessary to re-establish the proper neutron-to-proton ratio.

In the year 1939, a German radio-chemist, Otto Hahn, with his co-worker, Fritz Strassman, studied the effect of the neutron bombardment of uranium atoms, expecting to observe the formation of uranium isotopes with atomic weights higher than that of ordinary uranium. To his great surprise, Hahn found that the sample of uranium bombarded by neutrons contained radioactive atoms of a much lighter element, barium. The mystery of this discovery was soon cleared up by two German physicists, Lise Meitner and Otto Frisch, who suggested that in Hahn's and Strassman's experiments the nuclei of $U^{238}$ were split by incident neutrons into two nearly equal parts:

$$_{92}U^{238} + {_0}n^1 \rightarrow {_{56}}Ba^{145} + {_{36}}Kr^{94}$$

Since the barium and krypton atoms produced in this process possessed excess neutrons, as compared with ordinary stable atoms of the same atomic weight ($_{60}Nd^{145}$ and $_{40}Zr^{94}$), these so-called *fission* products emitted negative electrons, making them strongly radioactive. Frisch's and Meitner's interpretation of Hahn's and Strassman's experimental finding as the splitting of the uranium nucleus into two nearly equal parts opened new vistas in the field of nuclear physics. Instead of just "chipping off" small pieces of the bombarded nucleus, as was the case in all previous experiments, here was a real breakup of the central body of the atom, the *fission* of a large droplet of the nuclear fluid into two half-size droplets. Instead of just the few million electron-volts of energy observed in previous experiments on artificial nuclear transformation, uranium fission liberates 200 Mev per atom!

Detailed theoretical studies of the process of nuclear fission were carried out by Niels Bohr and John Wheeler (1911–present) and published in the September, 1939 issue of the *Physical Review*. This was the first and last comprehensive article on the theory of nuclear fission that appeared as open literature before the security curtain was drawn tight on that subject. According to Bohr and Wheeler, the fission of heavy nuclei resulting from the impact of a neutron is a resolution of a conflict between the opposing tendencies of nuclear (attractive) and coulomb (repulsive) forces acting in the atomic nucleus. When an incident neutron hits the nucleus and is absorbed by it, the excess energy communicated to the nucleus forces it to pulsate more or less as the water drops falling from the faucet do. There are two kinds of forces participating in that process.

1. The forces of nuclear surface tension attempting to bring the nucleus back to its original spherical shape.

2. Coulomb repulsive forces between the charges on the opposite ends of the ellipsoid attempting to break the nucleus into two halves.

In the previous chapter, we have seen that the nuclear model which considers the nucleus as a positively charged droplet of universal nuclear fluid leads us to conclude that for the lighter nuclei, the surface tension forces have the upper hand but that for the heavier nuclei, the electric forces become more and more important. Thus, we would expect that in the case of very heavy nuclei, the comparatively small deformation caused by the force of a neutron impact may result in a breakup (fission) of the original nuclear droplet into two halves.

## Fission Neutrons

In spite of the fact that each of the two fragments produced in the fission of uranium nucleus carries about 100 Mev of energy, these fragments are quite ineffective in producing further fission processes because the fission fragments carry a very high electrical charge and are consequently strongly repelled by the other uranium nuclei with which they may collide. Thus, the discovery of uranium fission would not contribute anything to the problem of the large-scale liberation of nuclear energy if it were not for a sceondary process that was found to accompany nuclear fission. It was discovered that, in addition to the two large fragments of the original nucleus, several extra neutrons are always emitted in the breakup. In the case of $U^{235}$, the average number of "fission neutrons" formed is 2.5

per uranium nucleus. These fission neutrons formed in the breakup of one uranium nucleus may collide with the surrounding uranium nuclei and produce more fission and still more fission neutrons. And, if the conditions are favorable, the breeding of fission neutrons goes *crescendo*, as does the breeding of rabbits on a rabbit farm or of fruit flies in a genetics laboratory. Thus, we get a *branching chain reaction*, and in practically no time at all the nuclei of uranium in a given piece of this material break up with the liberation of a tremendous amount of energy.

## *Fissionable Uranium-235*

As was mentioned above, natural uranium represents a mixture of two isotopes, $U^{238}$ and $U^{235}$, that are present in the relative amounts of 99.3 and 0.7 percent respectively. The study of these two isotopes under the influence of neutron bombardment had shown that the rarer isotope $U^{235}$ is much more fissionable than the more abundant $U^{238}$. Indeed, whereas $U^{238}$ nuclei do not break up unless the bombarding neutron has energy above 1.2 Mev, $U^{235}$ nuclei can be broken up by neutrons moving with much smaller velocities, and, in fact, the breaking-up probability increases with the decreasing velocity of incident neutrons.

In the range between the high energies needed to fission $U^{238}$ and the very low, or "thermal," energies favorable for fissioning $U^{235}$, neutrons are absorbed by $U^{238}$ without causing the latter to fission. Thus, the strong dilution of the active $U^{235}$ isotope by the inactive $U^{238}$ makes natural uranium just as useless for carrying out nuclear chain reactions as soaking wet logs are for building a campfire.

Indeed, most of the fission neutrons ejected in the breakup of $U^{235}$ nuclei in natural uranium will be captured by the much more abundant $U^{238}$ nuclei and thus will be taken out of the game.

Accordingly, in the early stages of nuclear energy development ("Manhattan Project"), much effort was spent on the separation of the active $U^{235}$ from the inactive $U^{238}$. Since the isotopes of a given element possess identical chemical properties, ordinary chemical separation methods could not be used in this case. The problem was finally solved by the development of the "diffusion separation" method, which was based on the fact that the lighter atoms of $U^{235}$ (and their various chemical compounds) diffuse faster through tiny openings than do the heavier $U^{238}$ atoms, and large amounts of "fissionable" uranium-235 were obtained this way.

## The Fermi-Pile and Plutonium

A good boy scout is supposed to be able to build a campfire even if the wood is soaking wet. This role of a good boy scout in the nuclear energy project was played by the Italian-American physicist, Enrico Fermi, who actually made the wet uranium logs burn. He was able to do so by utilizing the fact mentioned above, that the effectiveness of fission neutrons in producing the fission of $U^{235}$ nuclei increases when they are slowed down. If such slowing down of fission neutrons could be achieved, the presence of inactive $U^{238}$ would not make much difference, because very low-energy, slow neutrons are not absorbed by $U^{238}$ to any appreciable extent. To slow down the fast fission neutrons, it was necessary to use a *moderator*—i.e., some material from whose atoms the fast neutrons could bounce harmlessly and lose their energy. From considerations of conservation of momentum and energy, it can be shown that when a particle collides with another particle much less massive than itself, it is slowed down very little and loses scarcely any energy (recall the lack of effect that electrons have in deflecting $\alpha$-particles). At the other extreme, if a particle collides with another particle much more massive than itself, it bounces back with a speed and energy that are little changed. To be most effective, then, in helping the neutron lose energy by collision, the moderator atoms should be light atoms, comparable in size to the neutron, and should not absorb neutrons. To provide a moderator, it was decided to surround the pieces of uranium by carbon in the form of very pure graphite.

A large "pile" of graphite bricks with small pieces of natural uranium included in the structure was constructed in great secrecy under the grandstand of the University of Chicago Stadium, and on December 2, 1941, Professor A. Compton phoned to his colleague, Professor Conant of Harvard, the guarded message: "The Italian navigator has landed. The natives are friendly." This was quite correctly interpreted to mean: "Fermi's pile works successfully. The first successful nuclear chain reaction has been achieved."

In the pile, the fission chain reaction could be maintained in natural uranium, but the natural uranium was so highly diluted by carbon that high efficiency in energy production could not be achieved. Owing to the presence of inactive $U^{238}$, the chain reaction in the pile could not possibly develop into an efficient explosion, nor could it be very useful as a power source. So what good was the pile, except for demonstrating the purely scientific principle of the possibility of a self-maintaining nuclear reaction? Of course, the demonstration of a purely scientific principle is always of very great importance, but the pile was built at great expense in the midst of a perilous war when all expenditures were supposed to be judged on the basis of their military usefulness.

The Fermi-pile stood this acid test. Although the energy released in the fission of $U^{235}$ nuclei could not be utilized and was literally sent down the drain by means of the water-cooling system, a new fissionable element was produced inside the pile during its operation. The neutrons that were not used in the maintenance of the chain reaction in $U^{235}$ nuclei were captured by $U^{238}$ nuclei, producing the heavier isotope:

$$_{92}U^{238} + _0n^1 \rightarrow _{92}U^{239} + \gamma$$

Having an excess of neutrons, the nuclei of $_{92}U^{238}$ underwent two successive $\beta$-transformations, giving rise to elements with atomic numbers 93 and 94. These two elements, which do not exist in nature but have been produced artificially by human genius, were given the names *neptunium* and *plutonium*. The reactions following the neutron capture by $U^{238}$ can be written:

$$_{92}U^{239} \rightarrow _{93}Np^{239} + e-$$
$$_{92}Np^{239} \rightarrow _{94}Pu^{239} + e-$$

Being chemically different from uranium, the plutonium produced in the Fermi-pile can be separated and purified with much less effort than it takes to separate a light uranium isotope from the heavy one, and this element turned out to be even more fissionable than $U^{235}$. In fact, whereas $U^{235}$ gives rise to 2.5 fission neutrons, the corresponding figure for $Pu^{239}$ is 2.7 fission neutrons.

## Critical Size

When a single fission process occurs inside a given sample of pure $U^{235}$ or $Pu^{239}$, several fission neutrons are ejected from the point where the nuclear breakup took place. The average distance a fission neutron must travel through the material before it is slowed down to the point where it can effectively cause a fission reaction is about 10 cm, so that if the size of the sample in question is less than that, most of the fission neutrons will cross the surface of the sample and fly away before they have a chance to cause another fission and produce more neutrons. Thus, no progressive chain reaction can develop if the sample of fissionable material is too small. Going to larger and larger samples, we find that more and more fission neutrons produced in the interior have a chance to produce another fission by colliding with a nucleus before they escape through the surface, and for samples of a very large size only a small fraction of the neutrons produced in them has a chance to reach the surface before fissioning one of the nuclei. The size of the sample of a given fissionable material for which the percentage of neutrons giving rise to subsequent fission processes is high enough to secure a progressive chain reaction is known as the *critical size* for that particular material. Since the number of neutrons per fission is larger in the case of plutonium than in the case of $U^{235}$, the critical size of plutonium samples is smaller than that of $U^{235}$ samples because the former can afford larger losses of neutrons through its surface.

## Nuclear Reactors

As we have just seen, a sample of fissionable material smaller than the "critical size" is unable to carry on a nuclear chain reaction. If the size of the sample is *exactly critical*, the number of neutrons produced in each generation is the same as that produced in the previous one, resulting in steady nuclear energy liberation. The original Fermi-pile and its later modifications maintain nuclear reactions at the critical size level. It must be mentioned in this connection that the conditions of "criticality" are extremely unstable: a small deviation in one direction will result in the rapid extinction of fission neutrons and the cut-off of the nuclear chain reaction, whereas a deviation in another direction will lead to a rapid multipliction of the fission neutrons and the melting of the entire structure. Thus, the important problem in maintaining a steady chain reaction is that of regulating the rate of neutron production and of keeping the chain reaction from dying out or running away. This is achieved by using control rods made from neutron-absorbing materials (such as boron) which are automatically pushed in or pulled out from narrow channels drilled through the reacting fissionable material as soon as the rate of neutron production drops below or exceeds the desired level.

Many Fermi-piles are unsuitable for purposes of nuclear power production because of the high dilution of uranium by carbon; they should be considered rather as plants in which plutonium is produced. Because they can produce more fissionable material ($Pu^{239}$) than they consume ($U^{235}$), this type of pile is sometimes called a *breeder reactor*. For the purpose of nuclear power production, we can use controlled nuclear chain reactions in relatively pure fissionable materials, such as $U^{235}$ or $Pu^{239}$, which can be run at high temperatures. In the "swimming pool" reactor, in which several cylindrical containers filled with pure fissionable material are placed at the bottom of a large water tank, the water circulating through the tank carries away the heat produced in the fission process and also protects the observer from the deadly nuclear radiation. The water emits a ghostly blue glow as a result of what is called *Cherenkoff radiation*, after the Russion physicist who first analyzed the phenomenon in 1934. For this, and related work, Cherenkoff received the 1958 Nobel Prize in Physics. Cherenkoff radiation is produced by two separate effects that we have previously become acquainted with: the Compton effect and shock waves. We have spoken of the Compton effect before, primarily in terms of the scattering of an X-ray photon and its consequent loss of energy because of collision with an electron. In the reactor tank, high-energy γ-photons interacting with electrons send the electrons flying off at speed greater than the speed of light in the water. The result is similar to the bow wave of a ship traveling through water faster than the surface ripples can spread out: a shock wave (a bow wave for the ship) is formed. Analysis beyond the scope of this book shows that this electromagnetic shock wave will give rise to the blue Cherenkoff radiation.

In Figure 9-1*a* and *b* we give the schemes of two different types of *nuclear power reactors*. In both cases the block of fissionable material is perforated by long cylindrical channels for the passage of the "working fluid" that receives and carries out the heat produced in the fission process. Scheme (a) is known as a *closed-cycle nuclear reactor* since in it the working fluid (a molten light metal) is continuously circulating between the reactor and the "cooler," where the heat is used to produce water vapor for the operation of an ordinary steam turbine. This type of nuclear power reactor is installed, for example, in the "Nautilus," the first nuclear-powered submarine of the U. S. Navy. Figure 9-1*b* is an *open-cycle nuclear reactor* which is likely to become very useful for the propulsion of nuclear-powered jet planes. In this type of power reactor, the air coming in through the intake ducts in front of the airplane is heated to a high temperature while passing through the reactor and is ejected, in the form of a fast jet, through the nozzle at the rear.

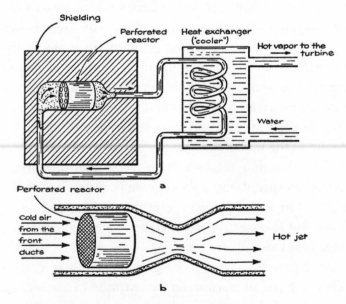

Figure 9-1. Two types of nuclear reactors for propulsion purposes: (a) a closed-cycle reactor and (b) an open-cycle reactor.

## Fission Bombs

If the sample of fissionable material exceeds the critical mass, the number of fission neutrons and the rate of energy production will increase exponentially with time, and the process will acquire an explosive nature. The principle of the fission bomb, or the "atomic bomb" as it is commonly called, consists of building up (assembling) a highly supercritical mass of fissionable material in such a short time period that the nuclear energy liberation that starts at the beginning of the assembly period does not develop to any important degree before the assembly job is finished. This can be accomplished in a simple way by inserting one subcritical piece of fissionable material into another subcritical piece as indicated in Figure 9-2. In order to perform the assembly process fast enough, we must shoot the inserted piece at a high speed from a gun muzzle, which earned for this assembly method the name of "gun gadget." There are also other more ingenious methods of bringing a given amount of fissionable material to supercritical size.

The energy liberation in the explosion of nuclear bombs is measured, according to established convention, in units known as *kilotons* and *megatons*, which refer to the weight of TNT (ordinary high explosive) that liberates the same amount of energy. One kiloton, i.e., the energy liberated in the explosion of 1,000 tons of TNT, equals $5 \times 10^{19}$ erg or about $10^{12}$ calories.

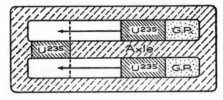

*Figure 9-2. The principle of the gun-type atomic bomb. The ignition of gunpowder (G.P.) shoots a U-235 cylinder along the axle and places it around another part of U-235, which forms the left end of the axle.*

## Thermonuclear Reactions

In experiments on the artificial transformation of light elements by electrically accelerated charged particles, one gets extremely low yields of nuclear reactions. The reason for this is that the charged nuclear projectiles, such as protons or $\alpha$-particles, rapidly lose their energy on the ionization of the material through which they fly, and only a very few of them (about 0.01 percent) have a chance to collide with a nucleus of the bombarded element before spending all their energy on tearing off atomic electrons. Thus, whereas the nuclear reactions that are caused by the high-energy projectiles produced in various kinds of high-voltage particle accelerators are extremely valuable for the study of nuclear properties, they are completely worthless for the purpose of large-scale nuclear energy liberation.

The most natural way of producing nuclear reactions in light elements on a large scale is the good old chemical method of raising the temperature of the reacting substances. Although in ordinary chemical reactions, a temperature of only a few hundred degrees may be quite sufficient to induce reactions between the colliding molecules, nuclear reactions require temperatures of many millions of degrees. At these temperatures, the atoms are completely dissociated into bare nuclei and free electrons, and the kinetic energy of thermal motion becomes high enough to overcome the electrostatic repulsion between the charges of the colliding nuclei. As in the case of nuclear bombardment by artificially accelerated particles, the energy of thermal motion does not need to exceed the height of the potential barrier separating the colliding nuclei, and penetration can occur by virtue of the tunnel effect for much lower energies. Whenever two colliding nuclei come into bodily contact, various types of fusion reactions may take place, and the nuclear energy flows out in a steady stream. These types of nuclear reactions, which are induced by the violent thermal collisions between the nuclei of the substance subjected to very high temperature, are known as *thermonuclear reactions*, and are responsible for the energy production in the interior of the sun and in other stors in the universe. The principal thermonuclear reaction responsible for the energy production in the sun and stars is a transformation of hydrogen into helium according to the formula:

$$4 _1H^1 = {_2}He^4 + 25.7 \text{ Mev}$$

However, even at the temperature of 20 million degrees in the solar interior, this reaction goes very slowly, liberating only $10^{-5}$ cal/gm/sec. The reason why our sun is so hot is that this meager heat produced by the thermonuclear reaction accumulates within its giant body before it has a chance to be radiated from the surface.

In man's search for thermonuclear reactions, he uses the faster-burning, heavier isotopes of hydrogen, deuterium ($_1H^2$ or D), which occurs in nature, and tritium ($_1H^3$ or T), which is nearly non-existent in nature and must be produced by nuclear reaction processes.

All possible reactions between these two isotopes are summarized below:

$$_1D^2 + _1D^2 \rightarrow _2He^3 + n + 3.25 \text{ Mev}$$
$$_1D^2 + _1D^2 \rightarrow _1T^3 + _1H^1 + 4 \text{ Mev}$$
$$_1D^2 + _1T^3 \rightarrow _2He^4 + n + 17.6 \text{ Mev}$$

These reactions go very fast even at "moderately low" temperatures. In Figure 9-3 we give the calculated rate of energy production in the D-D reaction. A similar curve can be constructed for the D-T reaction, which would run somewhat above the D-D curve. We see that, while at $10^{5\circ}$K this energy production is negligibly small, at a temperature of only $10^{6\circ}$K it rises to 1,000 cal/gm/sec, which corresponds to about 5 horsepower for each kg of material. At a temperature of $10^{7\circ}$K, the reaction rate rises to $10^{18}$ cal/gm/sec and all of the material will be consumed within a microsecond. This almost instantaneous release of nuclear energy leads to an explosion and is used in the construction of hydrogen bombs. However, to start the explosion, one has first to heat the material to the required high temperature, which can be done by using the ordinary uranium bomb as a starter. Notice here that the efficiency of such a thermonuclear bomb can be considerably increased by surrounding it with a layer of ordinary (cheap) uranium. In fact, although no fission chain reaction can be maintained in ordiary uranium, the numerous neutrons released by a thermonuclear reaction will cause individual fissions of uranium nuclei and add to the total energy release. Of course, such a design will result in the production of very large amounts of fission products, which will contaminate a wide area around the explosion and will be distributed by winds all over the globe.

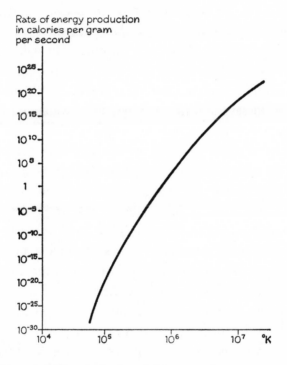

*Figure 9-3. The calculated rate of thermonuclear energy production in deuterium (at liquid density) for different temperatures.*

If, instead of having a violent explosion, we prefer to run a controlled thermonuclear reaction at a steady low rate, the physical conditions under which such a reaction can take place must be drastically changed. First of all, the reaction should be run at *extremely low gas densities*, since otherwise the pressure of the gas at the required temperatures of a few hundred thousand degrees will rise to millions of atmospheres and no walls will be able to contain it. Secondly, this rarefied gas *must somehow be kept away from the walls of the vessel*, since otherwise the process of heat conduction into the walls will rapidly reduce the temperature of the gas below the minimum value required for thermonuclear reactions. This can be achieved in several different ways, all of which are based essentially on the use of strong magnetic fields. At the very high temperatures required in this case, the deuterium gas in the tube will be completely ionized and will consist entirely of negatively charged electrons and positively charged deuterons. (This state of matter is described nowadays by the term "plasma.") We know that when an electrically charged particle moves through a magnetic field, it experiences a force perpendicular to the direction of its motion and to that of the field. This force compels the particles to spiral along the direction of the magnetic lines. Thus, by forming a strong axial magnetic field in a tube, we can effectively prevent free deuterons and tritons from coming close to the walls. If this can be achieved, the collisions between the particles spiraling along the tube are expected to result in D-D or D-T reactions with the release of nuclear energy and of large amounts of neutrons. Of course, in order to start such a process, the gas in the tube must first be heated to a very high temperature by some outside agent.

The second possibility consists in using magnetic forces caused by short but strong electric discharges through the tube. Two parallel electric currents flowing in the same direction are magnetically attracted toward each other so that, in the case of a sufficiently strong current, the gas (or rather the plasma) inside the tube will have a narrow jet along the axis. In contrast to the previously described method, the pinch-effect device operates in jerks, as an automobile engine does, but it has the advantage that the gas in the tube is automatically heated by the electric discharge, and no outside heating is needed. It has been estimated that a current of several hundred thousand amperes lasting for a few micro-seconds would produce a "pinch" strong enough to cause a thermonuclear reaction in deuterium. Work in the above-described directions is being carried out now in many laboratories of the world, and it is entirely possible that the problem of controlled thermonuclear reactions will be solved before this book comes off the press.

CHAPTER 10

# Mystery Particles

## *The Elusive Neutrino*

In addition to protons and neutrons, which form the nuclei of atoms, and electrons, which form their outer envelopes, physicists have discovered a whole array of other particles. These, although not permanent constituent parts of the nuclei, nonetheless play an important role in their properties.

Early studies of radioactive beta decay (the emission of an electron by an unstable atomic nucleus) led to the conclusion that something was wrong with the energy balance involved. Whereas alpha particles emitted by a given radioactive element always carry a well-defined amount of energy characteristic of that element, beta particles from one radioactive element show a wide energy spread, ranging from almost zero up to high-energy values. Since the total energy liberation in the transformation of one atomic nucleus into another is expected to be the same for all nuclei of a given kind, it was suspected that another particle must be coming out of the nucleus along with the electron and this carried away the missing balance of energy. This hypothetical particle, which must be electrically neutral and must be considerably lighter than the electron (we still do not know how light it is), received the name *neutrino*, which means "little neutral" in Italian. Their absence of electric charge and their extremely small mass allow neutrinos to penetrate thick material layers with the greatest of ease, and a heavy concrete wall is just as ineffective in stopping a beam of neutrinos as a chicken fence is in stopping a swarm of mosquitoes. In fact, it can be computed on the basis of theoretical considerations that, in order to stop a beam of neutrinos effectively, one would need a shield several light years thick! Thus, the neutrinos produced in various nuclear transformations could escape unobserved with their loads of energy, frustrating the physicists and causing discrepancies in the balance of the records of incoming and outgoing energy. But, whenever there is a suspicion of a new unknown particle, physicists are as good as Canadian Mounties in getting their man, and the nets were gradually drawn close around the elusive neutrino.

The first experimental evidence of the existence of neutrinos, which were originally introduced as purely hypothetical particles, was provided by the observation of the recoil of the nuclei from which the neutrinos were emitted. The unstable isotope of beryllium, $Be^7$, which can be produced artificially by means of nuclear bombardment, emits a positive electron and is transformed into the stable isotope of lithium, $Li^7$, according to the following equation, in which $\nu$ represents a neutrino:

$$_4Be^7 \rightarrow {}_3Li^7 + e^+ + \nu$$

Instead of this decay, however, another reaction can occur, in which the $Be^7$ nucleus is transformed into a $Li^7$ nucleus by capturing one of the electrons from the inner (K) electron shell of the atom:

$$_4Be^7 + (e^-)_{atomic} \rightarrow {_3}Li^7 + \nu$$

Indeed, the addition of a negative charge to a nucleus is equivalent to the loss of a positive charge. Since the captured negative electron belongs to the original unstable atom, all that happens here is the emission of a neutrino and, through the conservation of momentum, the recoil of the atom from which it came. Since the neutrino does not produce any visible track in the cloud chamber, it looks as if the $Be^7$ atom started suddenly to move by itself without any agent responsible for the move. It reminds one of an incident in an H. G. Wells story, "The Invisible Man," in which a self-respecting British bobby was suddenly catapulted forward by a kick in the pants while there was apparently nobody behind him. This phenomenon was actually observed in a cloud chamber containing unstable $Be^7$ atoms, and gave the first supporting evidence for the existence of neutrinos.

But the acid test of the neutrino hypothesis came in the attempt to stop the escaping neutrinos in their tracks. And, in spite of the almost incredible ability of neutrinos to make their getaway, physicists managed in 1955 to stop a few of them, thus finding unquestionable proof of their existence. F. Reines and C. Cowan of the Los Alamos Scientific Laboratory used for this purpose the collision process between neutrinos and protons in which the neutrino is expended to produce a positive electron and to transform the proton into a neutron:

$$p + \text{neutrino} \rightarrow n + e^+$$

These two scientists built a giant particle counter that registers neutrons as well as electrons and placed it near one of the nuclear piles at the Savannah River Nuclear Energy Project. The nuclear reactions taking place in the operating pile produce a tremendous number of neutrinos that stream out through a heavy shielding which holds back all other nuclear radiations. Although the chance of a neutrino hitting a proton and producing the above-mentioned reaction is only 1 out of $10^{30}$, some of these reactions do actually take place, resulting in the simultaneous appearance of a neutron and the accompanying positive electron. Thus, the uncatchable neutrino was finally caught and joined the company of well-established elementary particles.

## Mesons

The next member to enter the growing family of auxiliary nuclear particles was also born as the result of purely theoretical considerations. In 1935 a Japanese theoretical physicist, Hidekei Yukawa (known as "Headache" Yukawa to students who struggle with his mathematics), proposed a new particle which would account for the strong forces binding neutrons and protons together in the nucleus.

Probably the best way to picture a force of attraction between two bodies caused by the presence of a third body is to imagine two hungry dogs who come into possession of a juicy bone and are grabbing it from each other to take a bite. The tasty bone is continuously passing from the jaws of one of them into the jaws of the other, and in the resulting struggle, the two dogs become inseparably locked. Yukawa's idea was that attractive forces between the nucleons (the collective name for neutrons and protons) are due to a similar struggle for the possession of that new tasty particle. That new particle could be electrically neutral or could carry a positive or negative charge.

The binding energy between two nucleons, due to the periodic exchange of Yukawa's new particle between them, is $hv$ (as it is usual in all oscillation processes) where $v$ is the frequency of exchange. Since, according to the experimental data, this energy is about $10^{-4}$ erg, we conclude that

$$v = \frac{10^{-4}}{10^{-27}} = 10^{23} \text{ per second}$$

Neutrons and protons must have a very large appetite for this Japanese food if they fight so intensively for its possession!

According to Yukawa's theoretical considerations, the new particles must have a mass intermediate between that of protons and that of electrons, so they received the name *mesons* (from the Greek *mesos* meaning "between").

Two years after the introduction of these purely hypothetical particles for the explanation of nuclear forces, mesons were actually observed in cosmic rays by an

American physicist, Carl Anderson. The so-called primary cosmic rays bombarding the atmosphere of our planet are streams of extremely high-energy protons and a few other heavier positively charged nuclei that are probably accelerated by electromagnetic fields in interstellar space. The energies of these particles range from comparatively low values to thousands of billions of electron-volts. Colliding with the nuclei of atmospheric oxygen and nitrogen at the outer fringes of the atmosphere, these primary cosmic ray particles produce various kinds of penetrating radiations, including high-energy $\gamma$-quanta and streams of negative and positive electrons; in fact, as was mentioned earlier, positive electrons were first discovered in cosmic rays. Observing the tracks formed by cosmic ray particles in a cloud chamber placed between the poles of a strong magnet, Anderson noticed that the trajectories of some of the particles, both positively and negatively charged, were bent by a magnetic field more than would be expected in the case of fast protons but considerably less than should be the case with electrons. From the observed magnetic deflection, Anderson estimated that this new kind of particle was about 200 times heavier than an electron, which was in agreement with Yukawa's theoretical prediction. The behavior of the new particles, however, in their reluctance to react with nucleons, made it very doubtful that these were the predicted exchange-force particles. Ten years later the British physicist, C. F. Powell, demonstrated that there are two kinds of mesons: the $\pi$-meson (called "pion") which is produced at the upper fringes of the atmosphere by primary cosmic rays, and the $\mu$-meson (or "muon") into which the pion spontaneously decays on its way down in about $10^{-8}$ sec after being formed.

$$\pi^{\pm} \rightarrow \mu^{\pm} + \text{neutrino}$$

Muons, or $\mu$-mesons, have been detected by cloud chambers in relatively large numbers at the surface of the earth, and investigations have shown them to have a half-life of $10^{-6}$ sec, decaying into an electron and two neutrinos, according to the equation:

$$\mu^{\pm} \rightarrow e^{\pm} + 2v$$

According to Einstein's theory of relativity, $3 \times 10^{10}$ cm/sec is the absolute speed limit for any material particle. Even at this speed, which is the speed of light, the muon would require $10^{-3}$ sec to reach the earth's surface. If the mean lifetime of a muon is only $10^{-6}$ sec, how does it manage to survive the journey?

The answer to this question comes from the same relativity theory that enforces the speed limit. From our point of view, the watch of a swiftly moving observer will lose time; relative to us, the muon is traveling at enormously high speed, and the "watch" that times its life appears to us to be running slowly. In fact, the cosmic ray muons travel at a speed so near the speed of light that time is expanded by a factor of many thousands, and what we measure as a thousandth of a second will be to the muon much less than its lifetime of a millionth of a second. Besides explaining the survival of the muon, the above argument serves as another strong proof of the theory of relativity. Both pions and muons can carry either a positive or a negative electric charge ($\pi^+$, $\pi^-$, $\mu^+$, $\mu^-$), and in addition there also exist neutral pions ($\pi^0$). All of these new particles, as well as the positive and negative cosmic ray electrons ($e^+$, $e^-$), form a sequence somewhat similar to the sequence of the radioactive elements. It is now established that the primary high-energy (positive) protons entering the outer fringes of the atmosphere also give rise to neutral pions. Neutral pions possess a very short lifetime (about $10^{-16}$ sec) and, in spite of their high velocity, break up into two $\gamma$-quanta:

$$\pi^0 \to \gamma + \gamma'$$

long before reaching the surface of the earth. Positive, negative, and neutral pions interact very strongly with atomic nuclei and are apparently the particles introduced hypothetically by Yukawa for explanation of nuclear forces.

For the study of pions and their decay, photographic equipment attached to large balloons must be sent high into the stratosphere. Since cloud chamber equipment is too bulky and heavy to be sent up in balloons, cosmic ray researchers have developed a new method for photographing the tracks of cosmic particles at high altitudes. Instead of using the ionizing properties of fast charged particles passing through humid air, the new method is based on the fact that these particles affect the grains through which they pass when they travel through a fine-grained photographic emulsion. When the photographic plate is developed, it shows dark streaks that correspond to the trajectories followed by the particles. After the development of the Bevatron and Cosmatron, it became possible to produce pions in the laboratory and this considerably accelerated the progress of the study of their properties.

## More and More Particles

Following the discovery of pions and muons, other particles began to turn up. They appeared in cosmic ray studies and in experiments with new high-energy particle accelerators. Some of these particles (the K mesons) are intermediate in mass between electrons and nucleons, while other ($\Lambda$, $\Sigma$, $\Xi$ particles) are more massive than nucleons and are known as *hyperons*. Table 10-1 lists all the particles reported at the date of the conclusion of this manuscript, giving their masses, lifetimes, and modes of decay.

**TABLE 10-1 THE PROPERTIES OF THE ELEMENTARY PARTICLES OF MATTER**

*(BOLD TYPE INDICATES THE PARTICLES KNOWN BEFORE 1930)*

| Name and symbol | Mass (in electron masses) | Mean lifetime (in seconds) | Decay Scheme |
|---|---|---|---|
| **Xi $\Xi^{\pm}$** | 2,585 | $10^{-10}$ | $\Lambda^0 + \pi^{\pm}$ |
| Sigma $\Sigma^{\pm}$ | 2,330 | $10^{-10}$ | $n + \pi^{\pm}$ |
| Lambda $\Lambda^0$ | 2,182 | $2.7 \times 10^{-10}$ | $p + \pi^-$ or $n + p^+$ |
| Neutron n | 1,838.6 | $10^3$ | $p^{\pm} + e^{\pm} + \nu$ |
| **Proton $p^{\pm}$** | 1,836.1 | stable | |
| K-meson $K^{\pm}$ | 966.5 | $10^{-8}$ | $\pi^{\pm} + \pi^0 + \pi^0$ etc. |
| K-meson $K^0$ | 965 | $10^{-10}$ | $\pi^0 + \pi^0$ or $\pi^+ + \pi^-$ |
| Pion $\pi^{\pm}$ | 273.2 | $2.6 \times 10^{-8}$ | $\mu^{\pm} + \nu$ |
| Pion $\pi^0$ | 264.2 | $10^{-16}$ | two gamma rays |
| Muon $\mu^{\pm}$ | 206.7 | $2.2 \times 10^{-6}$ | $e^{\pm} + 2\nu$ |
| **Electron $e^{\pm}$** | 1 | stable | |
| Neutrino $\nu$ | 0 | stable | |

We do not know whether all the particles listed in the table are really elementary or whether some of them are formed by the combination of others. The experimental and theoretical studies of this problem represent the frontier of today's physics, and the solution will lead to a much deeper understanding of the nature of the physical world we live in.

## What Lies Ahead?

As described at the beginning of this book, ancient Greek philosophers, who conceived the idea of ultimate structural units of matter, recognized four kinds of atoms: those of stone, of water, of air, and of fire. The subsequent development of chemistry brought the number of different atomic species to 92, but up to the end of the nineteenth century, chemists and physicists were still subscribing to Democritus' dogma that the atoms of various chemical elements cannot be divided into still smaller parts. The work of Sir J. J. Thomson, Lord Rutherford, and other scientists who were digging into the atomic interior, proved, however, that the indivisibility of atoms is a myth and that, indeed, the atoms and their nuclei are extremely complex systems built from much smaller units which we now call elementary particles. First, the number of *elementary particles* was limited to two—*protons* and *electrons*—later joined by *neutrons*, which were considered to be just ordinary protons that had lost their electric charge. But soon hordes of new particles with the claim of elementary began showering down on the heads of the physicists. They were *neutrinos*, much lighter than an electron; several kinds of *mesons*, with masses between that of an electron and a proton; and *hyperons*, which exceed the proton mass. To each of these particles corresponds an *anti-particle* which can be annihilated in an encounter with a "normal" particle, the entire mass of both particles being turned into the quanta of radiant energy. More recently, physicists got into trouble with what can be called *mirror particles*. Everybody knows that, looking at a mirror image of the right hand, one appears to see the left hand, and vice versa. Since both right and left hands actually exist in reality, the "through-the-looking-glass" world coincides with the real world. Until recently, it was believed that the same is true in the world of elementary particles, that one can always find a real particle identical with the mirror image of another one. This dogma was known as the *conservation of parity*, a term referring to the symmetry properties of the equations describing physical phenomena. In 1957, however, two Chinese-American theoretical physicists, Chen Ning Yang and Tsung Dao Lee, concluded that, although conservation of parity undoubtedly holds for such phenomena as the emission of electromagnetic waves, it may not hold for the decay of elementary particles. Consider, for example, the transformation of a neutron into a proton with the emission of an electron. Protons and neutrons possess what is known as *spin* which may be interpreted as the result of their rapid rotation around the axes, and the emission of electrons is expected to take place along this rotation-axis. How is the direction in which the electron flies out of a neutron correlated with the direction of a neutron's rotation? One possibility would be that there is no correlation, that the electron could fly with equal probabilities in either of the two directions (i.e., either from the North Pole or from the South Pole, if we compare the rotation of a neutron with the rotation of the earth around its axis).

If we have two neutrons (a and b in Figure 10-1a) spinning in the same direction, say clockwise, but emitting electrons in opposite directions, their mirror images will appear to us as a' and b' in the same figure. But b' is identical with a turned upside down, and a' is identical with upside-down b. Thus the parity principle is fulfilled.

But if we assume that an electron is always emitted in only one definite direction in respect to the rotation of the neutron (only through the North Pole in our terrestrial analogy), the mirror image (Figure 10-1b) does not correspond to any real case of neutron decay and the parity principle is violated. Direct experiments with the beta decay of radioactive nuclei (which is due to the decay of neutrons in their interior) and also with the decay of mesons, have proved, however, that that is exactly what happens. Thus, it seems, that the world of elementary particles is lopsided, with one half of it missing, and nobody knows why.

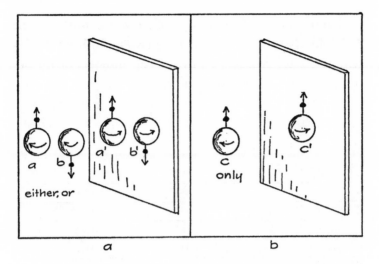

*Figure 10-1. Conservation (a) and nonconservation (b) of purity. In (a) the mirror images of a and b can be transformed into b and a by simply turning them upside-down. In (b), where the emission of an electron is coupled with the rotation of the neutron in a unique way, the mirror image does not exist in the real world.*

The world of elementary particles occupies the attention of physicists today, just as the world of atoms occupied it a few decades ago. Are these "elementary particles" really elementary in the good old Democritian sense of the word? Why do they have these particular masses, and why do they go through these particular transformations, with these particular rates? Will it be possible in the future to reduce the present fast-growing number of elementary particles to only two or three, as the 92 chemical elements were reduced to protons, neutrons, and electrons? Between the masses of elementary particles, we seem to find certain relations similar to those existing between the atomic weight of isotopes. Indeed, if we choose as a unit mass the 137-fold mass of an electron, the masses of other particles will be closely represented by integer numbers or integer numbers and a half (Table 10-2).

**TABLE 10-2**

| Name of particle | Measured mass in respect to electron | Mass divided by 137 |
|---|---|---|
| $\Xi$—hyperon | 2,585 | 18.88 $\rightarrow$ 19 |
| $\Sigma$—hyperon | 2,330 | 17.02 $\rightarrow$ 17 |
| $\Lambda^0$—hyperon | 2,182 | 15.95 $\rightarrow$ 16 |
| nucleon | 1,837 | 13.41 $\rightarrow$ 13.5 |
| $K^\pm$ and $K^0$—meson | 965 | 7.05 $\rightarrow$ 7 |
| $\pi^\pm$ and $\pi^0$—meson | 268 | 1.954 $\rightarrow$ 2 |
| $\mu^\pm$—meson | 206 | 1.504 $\rightarrow$ 1.5 |

The number 137 is not chosen at random but is a famous dimensionless physical constant equal to the product of quantum constant multiplied by the velocity of light divided by the square of elementary charge. In spite of numerous attempts, particularly those made by Sir Arthur Eddington, nobody knows why this ratio is 137 and not some other number.

One may say that the present opinion that elementary particles will really bear out their name (as the atoms did not) is due to our comparatively slight familiarity with their properties, and that all of them will be found in the future to be as complex as grand pianos. It may also be that this will not be the end of the road and that years later much smaller "subelementary" particles will be discovered.

There is no way to predict the future, and the question whether Democritus' original philosophical concept of indivisibility was right or wrong will never be answered by empirical means. But somehow many scientists, including the author, feel happier with the thought that, in the study of matter, "things will come to an end" and that the physicists of the future will know all there is to know about the inner structure of matter. And it also seems quite plausible that the elementary particles of modern physics really deserve their name, because their properties and behavior appear to be much simpler than could ever be said about the atom.

# PART 2

# Matter and Energy

# CHAPTER 11

# Bodies in Motion
# (Dynamics)

## How Things Fall

Dynamics is concerned with the motion of bodies and the action of forces in producing or changing their motion.

Probably the most familiar dynamic phenomena are present in the motion of material bodies acted upon by the forces of gravity, be it the fall of an apple from a tree, the flight of a ball on a tennis court, or (in more recent times) the flight of artificial satellites or spaceships. The first extensive investigations of the laws of fall are due to the famous Italian scientist, Galileo Galilei (1564-1642), who was a professor at the University of Pisa. His first experiment with free fall was aimed to disprove the opinion prevailing at that time that heavier bodies fall faster than light ones. In order to do this, he climbed, according to a contemporary story, onto the upper platform of the famous Leaning Tower of Pisa, carrying with him two spheres, one of solid iron and another of wood, and dropped them simultaneously over the railing (Figure 11-1). In spite of a large difference in weight, the two spheres fell side by side and hit the ground at practically the same moment. By this simple experiment Galileo established an important fact: independent of their weight, all material bodies fall equally fast. The fact that a leaf falling from a tree lingers in the air much longer than fruit falling from the same branch is due to the resistance of the air which slows down the free fall of lighter bodies more than that of heavier

Figure 11-1. Galileo's experiment.

ones. In a standard classroom experiment, one may demonstrate this fact by using a long, evacuated glass tube containing a feather and a coin. When the tube is rapidly turned upside down, the feather drops to the lower end as fast as does the coin. One can simplify this experiment by dropping a paper dollar and a silver dollar (or any other coin of the realm) onto the floor from the same height. The coin will hit first. But if one crumbles the paper dollar and rolls it into a small ball, it will fall almost as fast as the coin.

## Dilution of Gravity

Realizing that free fall was too fast to be studied in detail by the means which were at his disposal (fast movie cameras—any cameras at all, for that matter—had not yet been invented), Galileo decided to "dilute" the force of gravity and slow down the motion of falling bodies by making them fall along an inclined plane.

Having thus slowed down free fall, Galileo could observe the distances covered by a moving body during various periods of time. Since the stop watch had not yet been invented, he measured time intervals by the amount of water pouring from a faucet into a graduated beaker. This way he was able to prove that, if during a certain short period the rolling ball (originally at rest) covered a certain distance $s$, by the end of the second time interval the distance from the starting point was $4s$, by the end of the third it was $9s$, by the end of the fourth $16s$, etc. In other words, the distances covered by the ball during the various time intervals were increasing *as the squares* of these time intervals.

It can easily be shown that such a dependence between the time of travel and the distance traveled corresponds to the case when the velocity is increasing in direct proportion to time, or, as we say, when the motion of the weights is *uniformly accelerated.*

On the Moon, the acceleration of gravity is considerably less than that on the Earth (only 167 cm/sec$^2$), because the Moon is much smaller than the Earth (only $^1/_{81}$ of the Earth's mass) and therefore attracts the objects on its surface with a much weaker force. A man who on the Earth can jump, say, 2 ft into the air would soar up 12 ft on the Moon. On the surface of the Sun, on the other hand, the force of gravity is almost 30 times larger than on the Earth $(27,440 \text{ cm/sec}^2)$, and a man on the Sun would be crushed by his own weight if he were not first burned and turned into thin gas.

Closely related to the acceleration of gravity is the notion of escape velocity, i.e., the velocity that must be communicated (vertically) to an object to enable it to break the bonds of gravity and to escape into space. According to a formula which will be discussed later in this book, escape velocity is given by the square root of the product of acceleration of gravity on the surface of a celestial body and its radius. For the Earth, the escape velocity is 11.2 km/sec, while for the Moon it is only 2.4 km/sec.

The figure for the escape velocity from the Earth pertains to the velocity which the rocket must attain after it has passed the thin layer of terrestrial atmosphere and entered into empty space. Since rocket are sent from the bottom of the atmosphere, allowance must be made for friction losses during the short period of passing the atmospheric layer. If the rocket emerges from the atmosphere with a velocity only slightly higher than 11.2 km/sec, it will lose almost all of this in struggling against gravity, and will continue to move at a crippling speed. Thus, rockets designed for space exploration must have considerable excess velocity, beyond that needed for merely escaping from the Earth's gravity field.

## Newton's Laws of Motion

The year Galileo died in imprisonment, to which he was subjected by the Court of the Holy Inquisition for his belief in the Copernican system of the world, a boy called Isaac was born into the family of an English farmer named Newton. To Isaac Newton (1642-1727) science owes great discoveries in the field of mathematics (he invented calculus), optics (he discovered that white light consists of a mixture of different colors), and mechanics. All his works on mechanics were collected by Newton in his famous book, *Philosophiae Naturalis Principia Mathematica* (*Mathematical Principles of Natural Philosophy*), published in London in 1686. He put as the foundation of all his arguments and calculations the three basic laws which are now known as Newton's Laws of Motion.

*Newton's First Law of Motion* states: Every body continues in its state of rest, or of uniform motion in a right (straight) line, unless it is compelled to change that state by forces impressed on it.

Properly speaking, this statement should be considered not so much as a law of nature as the definition of force. In fact, in everyday life we observe force applied to some material object—for instance, when we see a horse pulling a wagon, or a man straining to lift a load. But we can conclude that a force is applied to a planet revolving around the Sun or to an electron revolving around the atomic nucleus, only on the basis of the fact that these motions deviate from rectilinear motion with a constant velocity. In his critical remarks on Newton's *Principia*, the famous British astronomer, Sir Arthur Eddington (1882-

1944) jokingly changes the passage following the comma to: "...*except insofar as it does not do otherwise.*"

Similarly, *Newton's Second Law of Motion* defines the way of measuring the strength of a force. It states: The change of motion is proportional to the motive force impressed, and is made in the direction of the right (straight) line in which that force is impressed. In Newton's terminology, "motion" is defined as the product of the mass of a moving body and its velocity; today we call this quantity the *mechanical momentum*. Thus we can write:

$$\text{rate of change of } (mv) \sim F$$

where $\sim$ is the sign of proportionality (means "is proportional to"). If $m$ is constant, this can be written as:

$$m \text{ (rate of change of } v) \sim F$$
$$\text{or}$$
$$ma \sim F$$

where $a$ is the acceleration. Thus the acceleration of a moving body is directly proportional to the force acting on it and inversely proportional to its mass.

The validity of Newton's Second Law of Motion can be conveniently demonstrated by means of the ingenious device known as Atwood's machine (Figure 11-2). It consists essentially of a long vertical pole with a light pulley on the top and a collection of identical metal discs (weights) which can be piled in any desirable quantity on the supports attached to the two ends of a string that runs

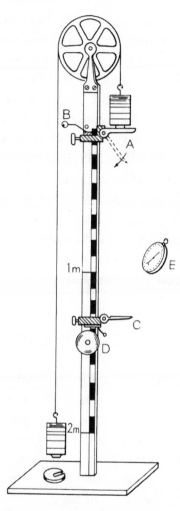

over the pulley. A platform *A* placed at the zero point of the vertical scale plotted on the pole supports the heavier pile of weights on the right side, and can be released by pressing the lever *B*. Another adjustable platform *C* can be placed at any height along the pole, and it rings the bell *D* when hit by the descending pile of weights. Using stopwatch *E*, we can measure rather accurately (at least for our purpose) the time between the release of the pile of weights from the upper platform and the moment that it hits the lower one. Suppose we place 51 weights on the right string and 49 on the left one. When the support is removed, the heavier pile of weights will descend in an accelerated way, but the acceleration will be much slower than that in the case of a free fall. Indeed, the acting force will be given by the difference between the two weight piles ($51 - 49 = 2$), while the mass to be moved will be equal

*Figure 11-2. Atwood's machine is used to measure acceleration.*

to their sum (51 + 49 = 100). According to Newton's Law, the acceleration will be 50 times smaller than that in the free fall. This arrangement accomplishes the same purpose as the inclined plane of Galileo, permitting a "leisurely" observation of the motion. Using Atwood's machine, one can easily demonstrate the law that the distances covered by a uniformly accelerated body increase as the square of time and find with good approximation the numerical value of *g*. It can also serve for a demonstration of the second Law of Newton. If one places 52 weights on the right and 48 weights on the left, the driving force will be that of four weights, whereas the total mass remains the same. Thus the acceleration will be twice as large as in the previous case which can easily be demonstrated by the experiment. On the other hand, putting 26 weights on the right and 24 on the left will result in the same force as in the first case acting on a half mass, and the observed acceleration will be correspondingly larger.

In the above discussion, we have always referred to the weight of material bodies and have used the word "mass" only in the colloquial sense. However, if properly defined, the notion of mass is more basic and general than that of weight. Suppose we take Atwood's machine to the Moon and repeat the experiments there. Since on the Moon gravity is considerably weaker than on the Earth, the gravitational pull on each individual weight will be correspondingly less. The inertial resistance of the weights against a change in their state of motion, however, will remain exactly the same, and as a result, all movements of the machine will be correspondingly slower.

Consider another experiment. Suppose we have two balls of equal diameter—one made of solid iron and the other of balsa wood—lying on a hard, smooth surface. If we try to lift the balls, it will take a much greater effort to lift the iron ball than to lift the wooden one, and we say that the iron ball is much heavier. If we push the balls along the surface, we will find that again the heavy one resists more stubbornly and requires a much stronger push to get rolling. This resistance of balls to the push has nothing to do with their weight or with the pull of gravity, since they are supported by the surface on which they rest, and the friction forces in this case can also be readily ignored. The iron ball resists our push more than the wooden one because of its larger *inertial mass.* Imagine now that we take the balls to the Moon and repeat the experiment. Because of the Moon's lesser gravity, we will be able to raise the iron ball without much effort, while the wooden ball will feel almost like a rubber balloon. If we try to push these balls along the surface, however, it will take just as much effort as it did on the Earth. Thus, although the weight of material bodies varies from place to place, their inertial mass, or simply their mass, remains always the same. This law makes it clear that we should use the (inertial) mass of an object rather than its weight for characterizing the amount of matter it contains. We do not need to construct a new universal unit for mass, since "standard kilogram," originally defined as the unit of weight, can also serve as the unit of mass. If we bring the standard kilogram from Paris to the North Pole and weigh it on spring scales, it will be heavier, while on the equator it will be lighter. On the Moon it will weigh much less than on the Earth, and on Jupiter much more. On the other hand, the inertial resistance of the standard kilogram to push will be exactly the same in Paris, on the Pole, on the equator, and on the Moon and Jupiter.

## *Unit of Force*

The Second Law of Newton can be used for the definition of the unit of force. Indeed, if in the formulae in the section "Newton's Laws of Motion" earlier in this chapter we replace the sign ~ with the sign =, the rational unit of force will be the force which, acting on a mass of 1 gm, results in the acceleration of 1 cm/sec$^2$. This unit of force is called a *dyne* (from the Greek *dynamis*, or power). According to the formula, the physical dimension of the force is:

$$|force| = |mass| \cdot |acceleration| = |mass| \cdot \frac{|distance|}{|time|^2}$$

or, in metric units,

$$dyne = \frac{gm \cdot cm}{sec^2}$$

Since, under the action of terrestrial gravity, a mass of 1 gm moves with the acceleration

$$g = 980 \frac{cm}{sec^2}$$

we must say that the force of gravitational attraction acting on mass $m$ on the surface of the Earth (i.e., its weight) is 980 dynes in the metric unit system; and so it will be used throughout this book. The relation between mass and weight becomes rather confused in the British-American unit system, where both are measured in pounds (lb) even though they possess different physical dimensions.

Also, while a given object has the same mass expressed in pounds in Edinburgh and in Gibraltar, its weight in pounds will be slightly different in these two locations.

## *Action Versus Reaction*

"Push me and I'll push you back!" are fighting words between schoolboys. Sir Isaac Newton puts it in slightly different words in his *Third Law of Mechanics:* To every action there is always opposed an equal reaction. This means that if there is an interaction between two material bodies, the force acting on one of them is exactly equal and opposite to that acting on the other. When a nurse pushes a perambulator along a smooth sidewalk, the perambulator pushes back at the nurse with exactly the same force. The reason that it is the nurse and not the perambulator who wins the competition is because her white Oxfords hold more strongly to the ground (higher friction) than the wheels of the perambulator (Figure 11-3a). If, however, the nurse gives a strong shove to the perambulator while standing on a slippery surface (especially if the perambulator is heavily loaded) the situation

*Figure 11-3. Newton's Third Law: The equality of action and reaction.*

will be different: the perambulator will move forward while the nurse will move backwards (Figure 11-3b).

Now, becoming more academic, let us consider two material bodies with the masses $m_1$ and $m_2$ attracting or repelling each other. According to the Third Law of Newton the two forces, $F_1$ and $F_2$, acting on these bodies will be numerically equal and opposite in direction ($F_1 = -F_2$). If the two bodies were originally at rest and the numerical value of the interaction force between them is $F$, their accelerations are $F/m_1$ and $-F/m_2$. As a result, after a period of time, their velocities ($v_1$ and $v_2$) will stand in the ratio.

$$\left( \frac{F}{m_1} \right) / \left( \frac{F}{m_2} \right) = m_2 / m_1$$

so that $m_1 v_1 = -m_2 v_2$ or $m_1 v_1 + m_2 v_2 = 0$. As was previously mentioned, the product of the mass of a body by its velocity is known as *mechanical momentum*, and the equation given here simply states that if the combined momentum of the two bodies was zero before the interaction, it will also be zero after interaction. When their combined momentum before interaction is not zero but has a certain numerical value $p$, no matter how the bodies interact with one another, their combined mechanical momentum will remain the same. This statement is known as the *Law of Conservation of Mechanical Momentum*. If *several* material bodies are placed in various points in space and not on a single line, the interaction forces between them will be represented by vectors, and it is the *vector sum* of these momenta which remains constant.

The principle of the conservation of mechanical momentum can be demonstrated by the very simple laboratory experiment shown in Figure 11-4. Two solid metal cylinders of equal diameter but different length are placed end to end on a support *S*. A small capsule filled with black powder is placed in a cavity between them with a wick sticking out. When the wick is lighted and the black powder blows up, both cylinders are thrown in opposite directions, with their velocities inversely proportional to their masses (2:1 in the drawing). Since according to Galileo's classical experiment all bodies fall equally fast no matter what their weight is, both cylinders will hit the surface of the table at the same moment. Since, however, the less massive cylinder will fly correspondingly faster, it will hit the table at a greater distance (twice the distance, in the drawing) than the more massive one. The points of impact can easily be noticed by the dents made by the cylinder's edges on the table's surface, or, better, on a layer of sand spread over it.*

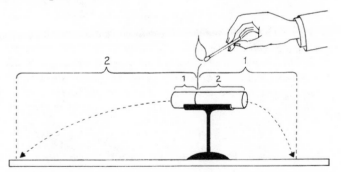

Figure 11-4. Lab test to show conservation of mechanical momentum.

---

* *Caution: Be careful not to use too much powder!* When the author and his laboratory assistant rehearsed this demonstration for a coming lecture, they used regular explosive fuses detonated by electric current. The cylinders flew far beyond the ends of the table, making dents in the walls of the classroom, and a dozen people rushed in from the neighboring rooms, attracted by the thunderous blast.

## Work and Energy

Let us turn now to two cases: 1) two forces applied to the ends of a solid bar supported at a certain point in between, and 2) two forces applied to the pistons of different diameters in two hydrostatically connected cylinders. We will assume here that the bar and the pistons have a negligibly small weight. These two cases with lever arms or area ratios, 1:2, are shown in Figure 11-5a and b.

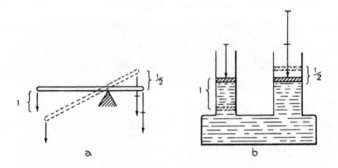

Figure 11-5. The equilibrium conditions for the two forces that are applied at the ends of a solid bar, and to the pistons in two hydrostatically connected cylinders. The directions of both forces are indicated by the arrows, and their relative strengths by the length of the arrows. Notice that the displacements of the ends of the bar, and of the pistons, are inversely proportional to the strengths of the two forces.

If the two forces satisfy exactly the conditions specified, both systems will be in the state of equilibrium and no motion will ensue, but if there is a very slight preponderance in one of the forces (say, the one on the left), the system will start to move: The left end of the bar and the left piston will move down and the right end and the right piston will move up. By inspecting Figure 11-5a and *b*, we find that the end of the bar or the piston acted upon by the double force moves only half the distance the other does. If the ratio of forces is 1:3, the displacement of the point of application of the larger force will be only one-third that of the other side. Thus we conclude that the product of an acting force and the displacement of the point of its application is the same for two nearly balanced forces that oppose one another.

This product is known in mechanics as the work done by the force, and we say that in the above cases the work done by two opposing forces is numerically the same. But, and it is an important "but," the force on the left did positive work, since it moved the object in the direction in which it was acting, and the force on the right did negative work, since it was "executing a strategic retreat under the enemy's pressure," as the military would say. Thus, if we ascribe to work done by forces a positive or a negative sign, as is done in algebra, the total work done by both forces in the above example is zero. This principle is the basis of a funicular railway, in which two carriages of equal weight and with about the same number of passengers are attached to the same cable, which passes over a pulley, so that the two carriages move in opposite directions up and down the side of a steep mountain. The only work done by the motor operating the funicular is that against friction and against the small difference in the weights of the passengers in the two carriages (the latter, however, averages zero in the long run).

The above definition of mechanical work is quite consistent with our common notion of the subject, at least in the case of crude physical work. We will all agree that it takes twice as much work to move twice as much furniture to the same floor of an apartment building or to move the same amount of furniture to an apartment located twice as high. So far we have been talking only about work done against gravity in lifting a certain weight to a certain height, but the definition of work as **the product of force by (times) the displacement of an object on which it is acting** is quite general. For example, when a horse pulls a loaded wagon along a level road, the amount of work it does is defined by the product of the force with which the wagon is pulled by the distance it is moved; but in this case, the work is done not against gravity (since the road is level) but against the friction of wheels against the axles and against the road surface.

If we choose as the unit of force 1 *dyne* (the definition of which was given earlier) and as the unit of length 1 cm, we get a metric unit of force known as 1 *erg*. Thus the physical dimension of erg is:

$$|erg| = |dyne| \cdot |cm| = \left|\frac{gm \cdot cm}{sec^2}\right| |cm| = \left|gm \cdot\right| \cdot \left|\frac{cm^2}{sec^2}\right|$$

—i.e., the same as the dimension of mass times the square of velocity. One erg is a very small unit of work, and, although it is always used in purely scientific studies, engineers prefer to use larger units such as "kilogram-meters" (in the metric system) or "pound-feet" (in the Anglo-American system) which represent the work necessary to lift the indicated weight to the indicated height. Since, as we have seen before, a force of gravity acting on 1 gm is equal to 981 dynes, the force acting on 1 kg is $9.81 \times 10^5$ dynes, and the work lifting 1 kg by (the distance of) 1 m = $10^2$ cm, is $9.81 \times 10^7$ ergs. The Anglo-American unit of work, i.e., pound-foot, is equal to 0.138 kg-m, of $1.36 \times 10^7$ ergs.

## *Potential and Kinetic Energy*

When an object is located at a certain height, it has the potentiality of producing work if and when it falls to the ground. The amount of work is known as the *potential energy* of the object, and is, of course, equal to the amount of work that had to be done to place the object in its elevated position. It has the same physical dimension as the work and is measured in the same units.

Returning now to our original experiments shown in Figure 11-1, suppose that the force on the left is considerably larger than the force on the right. In this case, "the orderly strategic retreat" will turn into "a rout in the face of the vastly superior enemy." Since the forces now are not balanced, the solid bar in the first example, as well as the pistons and the liquid in the second, will come into a state of accelerated motion and, gathering speed, will acquire the ability to produce a certain amount of mechanical work by virtue of that motion. **The ability of moving bodies to produce mechanical work is known as their kinetic energy** and can be expressed either in terms of the work done to bring them into that state of motion or in terms of the work they can do before coming to rest. We can investigate the situation by means of the simple experiment shown in Figure 11-6. A light carriage that can be loaded with a variable amount of heavy weights is rolled with a certain velocity along a table. A string attached to the carriage passes through a pulley of negligibly small weight and its other end is attached to a certain weight resting on the floor at the side of the table (a). When the string tightens up, the weight will be lifted from the floor and will reach a certain maximum elevation at the moment the

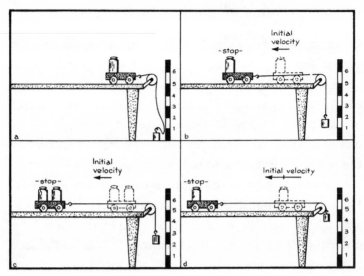

Figure 11-6. The work that can be done by a moving object increases in proportion to its weight and to the square of its velocity.

carriage comes to rest (b). If we double the load of the carriage, keeping its initial velocity the same, however, we will find that the weight will be raised twice as high (c). If, on the other hand, we use the same load but propel the carriage with a double velocity, the weight will be raised to four times the height recorded in the first experiment (d). Similarly, if we triple the load of the carriage, we will raise the weight to a triple height, and if we triple the velocity, we will find that the weight will be lifted nine times as high. Thus we must conclude that **the kinetic energy of moving objects is proportional to their total mass and to the square of their velocity.** We can therefore define kinetic energy as the product of these two quantities multiplied by a certain constant factor. The value of that factor depends on the choice of units of measurement and will be discussed at the end of this chapter. Being the product of the mass of a moving object by the square of its velocity, kinetic energy has the same dimension as potential energy and the work discussed earlier, and is measured in the same units.

In most examples of mechanical motion, there is a constant interplay between kinetic and potential energies. Thus, if we hold a ping-pong ball in our hand some distance above the floor, it has potential energy but no kinetic energy. If we release it, it falls faster and faster toward the floor, and when it reaches the floor, it will have no potential energy left; all of it will have been turned into kinetic energy. At the moment of impact with the floor, the ball will stop for a split second, and all its kinetic energy will be turned into the energy of the elastic deformations in its body. (In the case of an inelastic lead ball, the kinetic energy will be transformed into heat—by a process that we shall explain later—and it will not bounce.) The elastic energy is then changed back into kinetic energy and this sends the ping-pong ball up into the air, with the result that the kinetic energy is turned into potential energy. The ball will rise to approximately its original height. This process is repeated again and again until the friction forces gradually rob the system of its initial energy and the ball comes to a standstill on the floor.

The above discussion provides an interesting explanation of the fact that, although a man is several thousand times larger than a flea, a flea can jump to about the same height that a man can (Figure 11-7). Why is this so? If we use what we already know about work and energy, the explanation is simple. The body of any jumping animal is lifted into the air by the work done by its muscles. The larger the

animals, the larger are its muscles, and the greater is the total work they can produce. But how much greater? When a muscle contracts, the work it does is the product of the force with which it pulls and the length by

*Figure 11-7. A flea and a man can jump to about the same height.*

which it shrinks. Since muscle tissue is about the same in all animals, the tension produced per unit area of a muscle's cross section is also the same, so that the total force produced by a muscle is proportional to the square of the size of the muscle, or, what is the same, to the square of the size of the animal. In the case of a flea which is, say, 2,000 times smaller than a man, the cross-section of leg muscles, and also the force exerted by them, is $(2,000)^2$ times smaller than the cross-section of leg muscles in a man. The length by which a muscle contracts is a fixed fraction of its original length and therefore is proportional to the animal's size. Thus, the total work done by a contracting muscle is proportional to the cube of the size of the animal, which in the case of a flea is $(2,000)^3$ times smaller than in the case of a man. But the weight of the animal is also proportional to the cube of its size, and a flea weighs $(2,000)^3$ times less than a man. Using the formula, *work = weight × height*, we find that, since the work of the muscle and the weight of the animal change by approximately the same factor, the resulting height of a jump remains about the same. Of course, this is only an approximate relation, for there are good and poor jumpers among the animals of any one size.

## Power and Action

In the case of any device producing energy, be it a hydroelectric installation, a steam engine, or the old-fashioned working animal, we are interested, for the moment, only in its *power*, i.e., the rate at which its energy is produced. In purely scientific work, power is measured in *ergs per second* while engineers use *kilogram–meters per second* (= $9.81 \times 10^7$ erg/sec), *foot-pounds* per minute (= $2.26 \times 10^5$ erg/sec) etc. A popular technical unit of power is *one horsepower* (1 hp), equal to 75 kg-m/sec, or $3.25 \times 10^4$ ft-lb/min. There is, however, no "standard horse" in the Bureau des Pois et des Mesures in Paris, and this unit of power is simply defined in terms of other (previously given) purely mechanical units, and is roughly equal to what an old French horse can do in this way. (In electrical engineering, one uses *watts* (w) and *kilowatts* (kw), which are electrotechnical units which will be discussed in a later chapter.)

While the notions of force, energy, and power are quite familiar to everyone, the notion of *mechanical action* will cause most persons to raise their eyebrows, and means indeed something quite different from what its name may imply. In mechanics, the **action** is defined as **the product**

**of energy by time** (erg • sec), in contrast to the **power** which is **the ratio of the same two physical quantities**

$$\left( \frac{\text{erg}}{\text{sec}} \right).$$

Mechanical action, however, plays a very important role in various branches of physics. The famous Principle of Least Action, formulated by a French mathematician, P. L. M. de Maupertuis, a contemporary of Newton, states that all material bodies move through space from one point to another in such a way that the total action computed from the start to the finish is smaller than it would be if the body moved along *any other* trajectory and with *any other* velocities. In modern physics, the notion of action plays an important role in the quantum theory, permitting us to calculate the motion of electrons around atomic nuclei (as will be discussed later in this book).

# CHAPTER 12

# Vibrations and Waves

## *Surface Waves*

"If you are dropping pebbles into a pond and do not watch the spreading rings, your occupation should be considered as useless," said the fictional Russian philosopher, Kuzma Prutkoff. And, indeed we can learn much by observing these graceful circles spreading out from the punctured surface of calm water. When one of the waves encounters an obstacle such as, for example, the wall of the pond, it is reflected backwards as shown in Figure 12-1. The reflected wave looks as if it had been caused by a pebble dropped in the water at an exactly opposite point on the other side of the pond's wall. Thus, the wall of the pond acts as a mirror in respect to the surface waves, and, indeed, an optical mirror is based on the same principle except that it reflects waves of light instead of water waves.

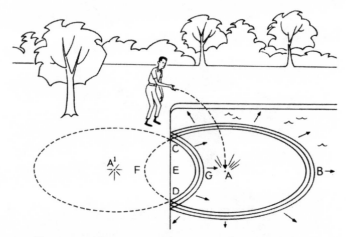

*Figure 12-1. The wave caused by an object thrown into a pool will be reflected off the side of the pool. This reflected wave will look as if it had been caused by an object tossed in at an exactly opposite (equidistant) point on the other side of the edge of the pool—provided, of course, that there was water on that side, too, and that the pool's edge did not interfere with the propagation of the wave from that side. (AB=AF; $A^1E=EA$; EF=GE)*

A propagating wave can be characterized by its *period, frequency,* or *wave length. Period* is defined as the time interval during which the source emits one single wave or during which a propagating wave passes any given point on its way. *Frequency* is the number of waves emitted by the source per unit time* or the number of individual waves which pass during a unit time through any given point. These two quantities are related by an obvious formula:

$$\text{frequency} = \frac{1}{\text{period}}$$

*often referred to as "per unit *of* time." Various technical terms of measurement drop the "of."

Thus, if a wave has a period of 1/100 sec, 100 waves will pass by during a 1-sec interval. The *wave length*, measured from crest to crest or from hollow to hollow, is related to the period (or frequency) of the wave and also to the velocity of its propagation. The wave velocity, which is the distance through which the wave motion spreads out in a unit time, is obviously equal to the number of waves emitted during that unit of time multiplied by the length of individual waves:

$$\text{wave velocity} = \text{frequency} \times \text{wave length}$$

or, in terms of the period:

$$\text{wave velocity} = \frac{\text{wave length}}{\text{period}}$$

For a more detailed study of the propagation of surface waves, it is more convenient to move from the pool to the laboratory and to use the arrangement illustrated in Figure 12-2a. It consists of a dish filled with water or mercury and an electrically driven elastic bar which operates on much the same principle as an ordinary electric bell. To the end of the bar are attached two vertical needles that barely touch the surface of the liquid. When the bar vibrates, the needles periodically disturb the surface of the water and send out two sets of concentric circular waves. The overlap of these two sets of waves produces a phenomenon known as *interference*, which is demonstrated in the photograph in Figure 12-3. The surface of the liquid breaks up into a number of narrow segments of alternately disturbed and calm water. The explanation of this phenomenon is given in Figure 3-7, which shows two waves propagating from the

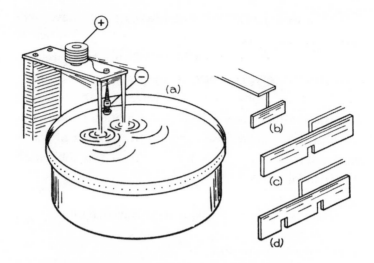

*Figure 12-2. Two electrically operated needles produce waves on the surface of a water tank.*

*Figure 12-3. The interference of two sets of waves propagating from two points. Courtesy Ealing Corp.*

points $O_1$ and $O_2$ on the left to a screen on the right. Since the distances $O_1A$ and $O_2A$ are equal, the two waves will travel to point $A$, with the ridges and hollows of the other. We say that the two waves are in phase with one another, and it is apparent that the resulting motion will be increased. On the other hand, if we select a point $B$ in such a way that the difference $O_1B - O_2B$ is equal to one-half of the wave length, the two waves will arrive out of phase, with the ridges of one overlapping the hollows of the other. It is clear in this case that the motion will practically cease. If we now select a point $C$ so that the difference $O_1C - O_2C$ is equal to the entire wave length, the two waves will be in phase again, and the motion will again be increased. We can proceed in this way selecting points $D$, $E$, etc. (not shown in the figure) that correspond to alternating in-phase and out-of-phase positions in which the surface motion will be alternately increased and obliterated. A similar interference pattern will exist, of course, in the lower part of the screen.

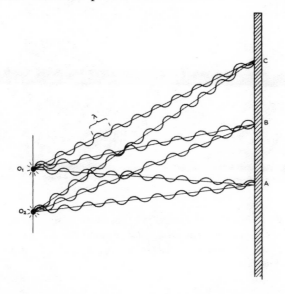

*Figure 12-4. The interference of two waves coming from the points, $O_1$ and $O_2$.*

Figures 12-5*a* and *b* show another experiment in which the two needles attached to the vibrating spring are replaced by a long vertical plate (see Figure 12-2*b*). In this case, we get what is known as a *plane wave* propagating across the surface of the liquid. If we place in its way a "breakwater" plate with a single opening in it (see Figure 12-2c), the wave will pass through and produce a pattern (shown in Figure 12-5a) that is similar to the pattern produced by a single oscillating needle. The pattern resulting from two openings in the "breakwater" plate (see Figure 12-2d) is reproduced in Figure 12-5b, and it again shows an interference phenomenon similar to that shown in Figure 12-3. The wave phenomena shown in Figure 12-3 and Figure 12-5 will have an important bearing on our study of light waves in Chapter 15.

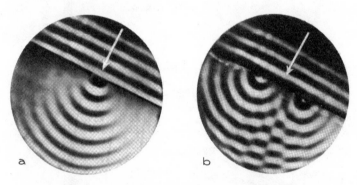

*Figure 12-5. The diffraction of a plane wave passing through a small opening in a breakwater (a), and the interference between two waves formed by two openings (b). Courtesy Ealing Corp.*

Before we leave the subject of waves on the surface of a liquid, we must say a few words about the forces that produce them. There are two kinds of forces—gravity and surface tension—that are responsible for the propagation of these waves, and which one predominates in a particular case depends on the length of the wave. In the case of ordinary ocean waves, the main operating force is the force of gravity. Since the water in the crest of the wave is elevated above the mean water level, when it comes down it pushes up the water in the trough next to it and thus makes the crest look like it is moving forward. But there is actually no horizontal or translatory motion at all. The particles of water execute periodic circular or elliptic motions as shown in Figure 12-6a and b. The deeper under the surface, the smaller the orbits described by the particles, and at the bottom the particles are motionless. Let us consider the lines $AA_1$, $BB_1$, etc. in Figure 12-6 as flexible boards attached to the bottom of any body of water. If all the boards move as indicated in the drawing, the water levels between them will rise and fall periodically and thus form a propagating wave.

In the case of waves operated by gravity, the velocity of propagation increases with the wave length; the long high waves in a stormy sea roll much faster than the ripple caused by a gentle breeze. Those of you who have been thrown rudely upon the shore by an incoming comber will probably object to the statement that waves have no translatory motion. Well, the waves that roll onto a beach can carry people and heavy objects for quite a distance, but contrary to the situation in the open ocean, the waves that come into shallow water become unstable, break up, and throw tons of water up onto the shore.

If we turn now to an analysis of short waves like those shown in Figures 12-3 and 12-5, we find that they are operated, not by the force of gravity, which is here quite small, but by the forces of surface tension. Since a plane surface has a smaller area than a wavy one, surface tension forces try to eliminate the crests and hollows in the same way that gravity forces do in the case of longer waves.

Although the short surface tension waves look much the same as the long gravity waves, they obey different laws of propagation; in particular, the velocity of these waves decreases as the wave length increases.

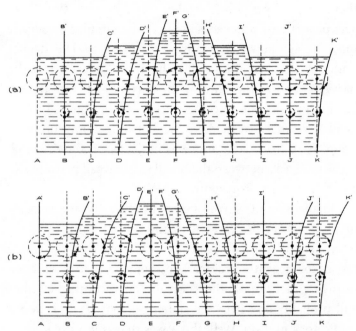

*Figure 12-6. The motion of water particles in a propagating wave. At the situations shown in (a), the crest of the wave is in F and the hollows in B and J. When the water particles move 45° along their trajectories, the crest is shifted to E and the hollows to A and I, giving the impression that the wave "runs" from right to left (b).*

There are other differences, also, between the two kinds of waves. Since long waves are operated by gravity and since their propagation results from the interplay between the weight of water elevated above the mean level and the mass of water to be moved, their velocity would remain the same even if the oceans were filled with a lighter liquid such as alcohol or a heavier liquid such as mercury. We have here the same situation as that in Galileo's experiment of dropping a light and a heavy body from a tower. Ocean waves on the Moon (if there could be any oceans there), on the other hand, would have a different propagation velocity than those on the Earth because the Moon's gravity is less. The properties of surface tension waves, however, largely depend on the substance in which they are propagating. Thus, if we substituted mercury for water in Figures 12-3 and 12-5, the waves would run considerably slower, but it would not make any difference whether the photographs were taken on the Earth or on the Moon.

Returning to the waves produced by a pebble in a pond, we may ask whether these waves are "long" waves operated by gravity or "short" waves operated by surface tension. As it turns out, in the case of water, the demarcation point between "long" and "short" waves is a wave length of about two centimeters. Since this length is just about that of the waves of the type shown in Figure 12-1 are operated by both gravitation and surface tension forces.

Although we ordinarily see surface waves only on the surface of water or some other liquid, such waves also occur on the surfaces of solids and gases. In an earthquake, for instance, although the main destructive force propagates right through the body of the Earth, some surface waves are also created that may be quite destructive over a comparatively small distance. To get an example of surface waves in a gas, we have to turn to the terrestrial atmosphere, the upper boundary of which, though not as sharply defined as the surface of the ocean, is still some kind of a boundary. When in the year 1883 the volcano Krakatoa exploded in the Dutch East Indies with an energy of thousands of 10-megaton hydrogen bombs, a wave of disturbance in the atmosphere traveled several times around the Earth. The pressure changes observed at that time at meteorological stations all around the globe were interpreted to be the result of a surface wave that propagated along the outer boundary of the terrestrial atmosphere.

## Waves in Solids

Elastic disturbances that propagate through solid bodies are of two different kinds: *transverse waves* and *longitudinal waves*. Suppose a horizontally suspended metal bar is struck by a hammer at one end in the direction perpendicular to its length (Figure 12-7*a*). The impact of the hammer will cause a deformation (bending) near the point of impact, and this deformation will propagate along the bar just as the surface wave did in the pond of water. If we continue to hit the bar with regularly repeated blows, we will produce a steady train of waves along it. Waves of this kind are known as *transverse* or *T-waves*, and the motion of the material particles is perpendicular to the direction of propagation.

If, instead of hitting the bar sideways, we hit it in the direction of its axis as shown in Figure 12-7*b*, we will produce a rather different kind of wave. The material of the bar will be compressed at the point of impact, and this compression will propagate along the bar without causing any sidewards motion at its surface. This type of wave is a *longitudinal* or *L-wave*, and the particles of metal move back and forth along the direction of the propagation. The two kinds of waves, *T* and *L*, will, generally speaking, move at different speeds since they depend on different mechanical characteristics of the material: the resistance to bending (or shearing) in the first case and the resistance to compression in the second.

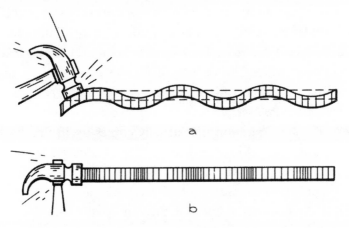

*Figure 12-7. Transverse and longitudinal waves in a solid bar.*

## Sound Waves

Since a fluid substance, whether it is a poorly compressible fluid such as water or a readily compressible fluid such as air, does not offer any resistance to the changing of its shape (such as bending or shearing), transverse waves cannot exist in a fluid. The only waves that can propagate through a fluid are longitudinal compression waves. These compression waves, as they propagate through the water of the oceans and the air of the atmosphere, play a very important role in all walks of life. They warn a convoy ship of the presence of an enemy submarine, a peaceful gazelle of the approach of a beast of prey, and a motorist of the onrushing engines of the fire department. They carry the love calls of birds and Indians (but not those of fish!), they bring high profits to music halls, and they help professors communicate their knowledge to classrooms full of students. These waves are, of course, those commonly known as *sound waves*. Because of the practical importance of sound waves, the branch of physics dealing with this comparatively narrow subject is highly developed and is known by the special name of *acoustics*.

The first question that usually comes up in this field concerns the velocity of sound. If we watch an artillery battery practice some distance away, we will see the flash of light a few moments before we hear the sound of the shot. And, we all know that the clap of thunder follows the lightning with a delay that depends on how far away the thunderstorm is. Since light propagates practically instantaneously (by our everyday standards, of course), the velocity of sound can easily be found by timing the lag between the flash of light and the roar of the shot and then measuring the distance between the observer and the artillery piece. In this way we find *that the velocity of sound in air under normal atmospheric conditions is 330 m/sec.* It was also found that the velocity of sound in air is not affected by any changes in air density but increases noticeably with the temperature.

We can try to calculate the velocity of sound by using dimensional analysis. Assuming that this velocity must depend on various powers of the temperature $\theta$, density $\rho$, and pressure $p$ of the medium through which sound waves propagate, we can write:

$$v = \delta \cdot \theta^x \rho^y p^z$$

where $\delta$ is a dimensionless number and $\theta$ is measured from the absolute-zero temperature ($-273°C$) as will be explained in the next chapter. Remembering that the dimension of $\rho$ is:

$$\frac{gm}{cm^3}$$

and of $\rho$ is:

$$|p| = \frac{|force|}{|area|} = \left|\frac{gm\ cm}{sec^2}\right| \cdot \left|\frac{1}{cm^2}\right| = \left|\frac{gm}{cm\ sec^2}\right|$$

we obtain:

$$\left|\frac{cm}{sec}\right| = |\theta|^x \cdot \left|\frac{gm}{cm^3}\right|^y \cdot \left|\frac{gm}{cm\ sec^2}\right|^z$$

This leads to the equations:

$$0 = x$$
$$0 = y + 2$$
$$1 = -3y - z$$
$$-1 = -2z$$

from which follows:

$$x = 0 \qquad y = -^1/_2 \qquad z = +^1/_2$$

so that our formula becomes:

$$v = \delta \sqrt{\frac{p}{\rho}}$$

Normal atmospheric pressure is about

$$10^6 \frac{\text{dynes}}{\text{cm}^2}$$

while normal atmospheric density is about

$$10^{-3} \frac{\text{gm}}{\text{cm}^3}.$$

Substituting these values in the above formula, we get

$$\delta \cong \delta \sqrt{\frac{10^6}{10^{-3}}} = \delta \cdot 3.16 \cdot 10^4 \text{ cm} / \sec = \delta\, 316 \text{m} / \sec$$

which agrees well with the experimental value if $\delta$ is not much different from 1.* The fact that $x = 0$ seems to indicate that $v$ is independent of the temperature, which would contradict the previously quoted experimental result.

But there is a trick. For air, as well as for any other gas, there exists a relation: $\rho \sim p\theta$ which, being substituted in the formula given above, gives us:

$$v \sim \sqrt{\frac{p}{\rho}} \sim \sqrt{\frac{\rho\theta}{\rho}} \sim \sqrt{\theta}$$

*In fact, the exact calculations show that $\delta = \sqrt{\frac{5}{3}} = 1.29.$

which is in good agreement with the above-stated experimental findings. In the case of liquid or solid materials, the situation is quite different. In water, sound waves propagate about 4.5 times faster than in air, and in steel the velocity is 15 times greater. In both cases the dependence of sound velocity on the temperature is very slight.

Sound waves differ from each other by their vibration frequency and wave length. Compression waves in air can have any frequency and any wave length, but when we speak about sound, we usually mean audible sound. The classic device for studying the vibration frequency or pitch of sound is a siren. A metal disc, containing a number of holes along its rim, is driven by a motor at varying speeds. A steel cylinder supplies the compressed air that comes out in a putt each time an opening in the rotating disc aligns with the end of the pipe that releases the compressed air. If the disc rotates comparatively slowly, we will hear: "puff, puff, puff…," not unlike the sound that comes from the exhaust pipe of an automobile when the engine is idling. As the rotation velocity of the disc increases, we will hear a musical tone that will sound first like a bassoon, then like a flute and, finally, like a piccolo. If we speed up the rotation of the disc still more, the sound will fade out and we will hear practically nothing. Knowing the rotation velocities of the disc in each case and the number of holes along its rim, we can easily calculate the number of "puffs per second" that correspond to the different sounds we hear (or don't hear). In this way we find that the lowest frequency which our ear accepts as a tone is somewhere around 20 oscillations/sec and the highest audible frequency is nearly 20,000 oscillations/sec. By remembering that sound propagates at the rate of 330 m/sec, we can calculate the wave lengths of audible sound and find them to range from about 1.5 cm to about 15 m.

## Supersonics and Shock Waves

It is simplest to discuss the problem of motion of a material object through a medium in which the disturbances caused by its motion propagate with lower velocity than that of the object itself, by considering a ship sailing over a smooth surface of the sea. The displacement of water caused by the advancing bow of the ship propagates in the form of surface waves, just as in the case of a stone dropped in water. If the velocity of these waves $v$ is higher than the velocity $V$ of the ship, they run ahead of it, and things looks as shown in Figure 12-8a. If, on the other hand, the ship moves faster than the waves produced by its advancing bow, the picture looks quite different (Figure 12-8b). The water surface in front of it remains undisturbed and we observe *two* wave fronts, propagating sidewise in a direction forming an angle $\theta$ with the course of the ship. It is easy to see that the angle $\theta$ is determined by the equation:

$$\sin\theta = \frac{V}{v}$$

and the higher the speed of the ship, the sharper the wave angle made by its bow. The velocity of surface waves in deep water produced by a sailing ship is of the order of a few meters per second, or in naval terminology just a few knots (1 knot = 0.51 m/sec). Thus it would be a very slow ship which moved more slowly than the wave it produced. Modern ships making 20 and more knots move much faster than waves; the pattern of water disturbance has the shape shown in the lower part of Figure 12-8.

*Figure 12-8. The shape of bow waves of a ship moving (a) slower than water waves, and (b) faster than water waves.*

In the case of objects flying through the air, the situation is quite different, since the velocity of compression waves produced by them (sound velocity) is 330 m/sec, or about

1,000 km/hr (620 mph). Until recently, such velocities could not be obtained either by projectiles or airplanes; the inhabitants of cities bombed during World War II could hear the whistle of falling bombs before they were hit by them. It is only recently that we have entered into the era of supersonic planes and missiles which hit you before you hear their approach. It is customary to express the velocity of fast objects moving through the air in their *mach number* (so called in honor of an Austrian physicist, E. Mach), which is simply the ratio of the object's velocity to the velocity of sound. While in the case of the subsonic vehicles we observe nothing but a turbulent wake dragging behind them—which is not much different from the wake behind a ship's stern—the supersonic vehicles show, apart from the wake, a sharp discontinuity of air pressure at their advancing front edge. These discontinuities of pressure moving along with the advancing object are stationary in respect to it and are therefore known as standing shocks. These standing shocks represent a serious danger in the flight of supersonic planes, and may cause considerable damage to their structural parts, resulting in fatal crashes.

When a plane or a guided missile moves through the air at subsonic speed, the resistance of air to its motion, known as *drag*, decreases slowly with its velocity, as indicated in Figure 12-9. When the velocity of the plane approaches the speed of sound, the plane needs a lot of additional energy to overcome the standing shocks, and the air resistance to further acceleration sharply increases. This resistance encountered in the transition from subsonic to supersonic speed is known as the *sonic barrier*. As soon as the velocity of the plane exceeds the velocity of sound, however, the increase of drag slows down and the resistance to further acceleration becomes more tolerable.

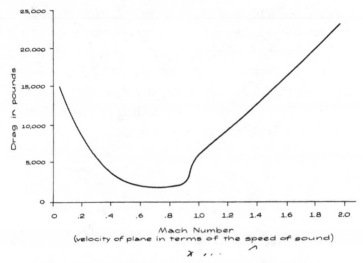

Figure 12-9. *The drag of a typical high-speed fighter-type airplane at 35,000 feet. Courtesy Convair, San Diego, Calif.*

## Shock Waves Caused by Explosions

Shock waves caused by explosions are similar to the standing shocks which accompany supersonic vehicles. Indeed when an ordinary high-explosive bomb is detonated, its initially solid material (TNT, RDX, etc.) turns into a hot gas so quickly that in the first instant it occupies the same volume as in the solid state. The enormous pressure tears the metal shell and the hot gases expand into the surrounding medium (air or water) at supersonic speed, driving the compressed material ahead of them (Figure 12-10a). As the expansion proceeds, the gas formed by the explosion gradually cools down and its pressure decreases. At a certain stage the compressed layer of medium separates from the slowed-down gases and the discontinuity of pressure (shock wave) propagates freely through the surrounding medium, causing damage to any object it encounters on its way (Figure 12-10*b*). At still larger distances, the advancing shock wave degenerates into sound waves, and people standing at a safe distance away hear only the roar of the explosion.

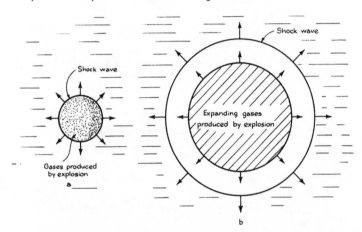

Figure 12-10. *The formation of a shock wave in an explosion.* (a) *At the early stage of the explosion, the gases formed by the exploded material push through the surrounding medium;* (b) *at the later stages, the discontinuity of pressure separates from the expanding gases and propagates freely through the surrounding medium.*

In the case of uranium and hydrogen bombs, the situation is somewhat different. Nuclear energy is liberated in a comparatively small volume of fissionable or fusionable material, mostly in the form of highly penetrating high-frequency radiation (γ-rays) which spread with the velocity of about 1 percent of the speed of light into the surrounding medium, heating it to the temperature of about 300,000°C * until all the radiation energy is spent. This process lasts only a fraction of a millisecond (one-thousandth of a second) and results in the formation of a luminous sphere called the *fireball*, the radius of which depends on the energy liberated by the bomb (about 45 ft for the 20-kiloton bomb). This ball of fire corresponds to the original sphere of hot gases formed in ordinary explosions before they began to expand.

The next step in nuclear explosions is analogous to that in the ordinary (chemical) ones. Hot air (or water vapor in underwater shots) begins to expand into the surrounding medium, forming a shock front which, at a certain stage, separates from the ball of fire and propagates freely in all directions. Figure 12-11 is a photograph of the first uranium bomb explosion (Trinity test) at Alamogordo on July 16, 1945. It was taken 15 milliseconds after the detonation, at which time the radius of the fireball increased from the original 45 ft to about 300 ft, and its surface temperature dropped to about 5,000°C. At this stage the shock front had just separated from the fireball and was now propagating in the free air. The darker region around the lower part of the fireball is known as *mach front* and represents the part of the original shock front reflected from the ground. The cottonwood-looking ring at the bottom represents a dirt cloud raised by the shock wave from the surface of the desert. So much about explosions and shock waves.

---

*Remember that the temperature of the Sun's surface is only 6,000°C.

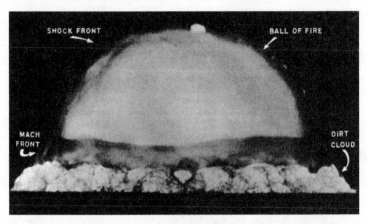

Figure 12-11.    The first nuclear explosion (A-bomb) at
Alamogordo, N.M., on July 16, 1945. Courtesy Atomic Energy
Commission.

# CHAPTER 13

# Heat and Temperature

The only way to explain to a person the meaning of the words "hot" and "cold," if he claims that he does not know what they mean, is to put him first into a tub filled with almost boiling water and then into one with ice cubes floating in it. Fortunately, such drastic treatment is never necessary since we all possess a well-developed sense of temperature. All physical bodies, such as a piece of solid material or a certain amount of liquid, respond to temperature changes in a simple way: they expand when the temperature rises and contract when it drops.

## *Absolute Zero*

When a gas, any gas, is heated from the freezing point of water to its boiling point, its volume increases by about one-third or, to be more exact, by $1/2.73$ of its original value. Since we have agreed to measure the temperature by the volume changes of gas and to divide the temperature interval between the freezing and boiling points of water into 100 equal parts, or *degrees centigrade* (°C), we say that 1°C corresponds to the volume change of gas by one 273rd part. If we now begin to cool the gas below the freezing point of water, it will contract by that fraction for each degree of cooling so that, at the temperature of 273°C below the freezing point, the volume of any gas should be expected to become zero. This point is known as the *absolute zero* of temperature, and the temperature counted from that point is known as *absolute temperature* (°abs.) or *Kelvin temperature* (°K). In Figure 13-1 we give a graphic presentation of volume changes in gases as the function of absolute temperature. As long as a gas remains a gas, the graph is a straight line passing through absolute zero, and the gas shows every intention of shrinking to zero volume at that temperature. These intentions are never exactly fulfilled, however, since all gases liquefy before they reach zero. Some do it sooner, some do it later, and helium does it last of all, at only about four degrees before reaching absolute zero. And, of course, as soon as a gas turns into a liquid, its volume decreases much slower and does not tend to zero any more. But, although no real gas actually goes to the end of the track, the notion of absolute zero temperature is very important in physics and can be, if desired, used in reference to an imaginary "ideal gas" which remains a gas no matter how much we cool it.

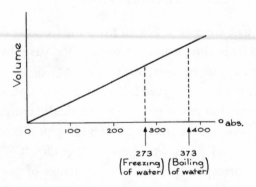

*Figure 13-1. The behavior of gas at low temperatures.*

## *Amount of Heat*

If we take a glassful of water at, say, 80°C, and mix it with an equal amount of water at, say, 50°C, we will find that the mixture will have a temperature of 65°C, i.e., just in between. But if we mix one glass of 80°C with two glasses at 50°C, we find that the temperature of the mixture will be only 60°C. This and similar observations can be interpreted in the following way: *Each material body contains in itself a certain quantity of what we call "heat," and the total amount of that heat increases with increasing temperature.* When we mix a glass of hot water with a glass of colder water, the excess of heat in the former is equally distributed between the water of both glasses. Since each volume of mixed water now has only half of the heat excess formerly existing in the first glass, the temperature of the mixture will drop halfway. In the case of hot water mixed with two glasses of cold water, the original heat excess is "diluted" by a factor of three and the temperature excess above the original cold water mark will be only one-third of the difference between 50°C and 80°C.

Having established the notion of the amount of heat, we can now define the unit for measuring it. In scientific measurements, we use the unit known as the calorie (cal), which is defined as *the amount of heat necessary to raise the temperature of 1 gm of water by 1°C.* A larger unit, the *kilocalorie* (kcal), is defined as 1,000 cal.

Different substances have different heat capacities, which are called *specific heats* (if they refer to unit mass)— that is, they require different amounts of heat to raise the temperature of one gram of the substance by one degree. It is interesting to notice that water has an exceptionally high heat capacity, so that the figures for other substances are, as a rule, considerably smaller than unity; for instance, the specific heat of alcohol is 0.232 and that of mercury 0.033.

## Latent Heat

When we place a teakettle on the fire, the temperature of the water gradually rises to 100°C, at which point the water begins to boil. But, once the boiling has started, the temperature stays at 100°C until the last drops of water are turned into steam. Although the heat is still flowing into the kettle from the flame, it does not make the water any hotter. What happens to that heat? The answer is, of course, that this heat is used to transform the water into vapor, and measurements show that to do it we must supply 539 cal/gm of water to be vaporized. This amount of heat is known as the *latent (hidden) heat of evaporation*, and is, of course, different for different substances. Thus, to evaporate 1 gm of alcohol and 1 gm of mercury, we need only 204 and 72 cal respectively. The heat absorbed in the evaporation of water plays an important role during hot weather in the cooling of our body through the process of skin perspiration. Indeed, one glass of water evaporated from the surface of our body removes enough heat to cool the entire body by a few degrees. If the weather is "sticky," with a high content of vapor in the atmosphere, the evaporation is considerably slower; a layer of water is formed on the skin, and we begin to sweat. Meteorologists use the same principle for measuring the relative humidity of air. The apparatus used for this purpose is known as a "psychrometer" and consists of two identical thermometers with the ball of one covered by a wet cloth. This thermometer, because of evaporation, shows a somewhat lower temperature, and from the difference between the two readings the weatherman can calculate the rate of evaporation and, consequently, the amount of humidity present in the atmosphere.

A similar phenomenon is encountered when water turns into ice. When the temperature of water comes down to 0°C and the first crystals of ice begin to form, the temperature remains at zero until all of the water freezes. The *heat of fusion of water* (i.e., the amount of heat that must be taken away from 1 gm of water at zero to freeze it, or be given to 1 gm of ice at zero to melt it), amounts to 80 cal. The heat of fusion of alcohol (which freezes at −114°C) is only 30 cal/gm, whereas for mercury (freezing at −39°C) it is only 2.8 cal/gm. To melt lead (at +327°C) it takes about 6 cal/gm, whereas in the case of copper (at +1,083°C), the figure is as high as 42 cal/gm.

## Mechanical Energy Versus Heat

In Chapter 11, we mentioned that the law of conservation of mechanical energy is challenged by the friction forces that gradually rob mechanical systems of their energy and eventually bring them to a standstill. On the other hand, we know that where there is friction there is always heat, be it in the two sticks of a Boy Scout trying to build a fire in the old Indian way or in the axles of a railroad car overlooked by the oil man. What is the relation between the mechanical energy lost to friction and the amount of heat produced by it? This question was answered in the middle of the last century by a British physicist, James P. Joule (1818-1889), in his famous experiment on the transformation of mechanical energy into heat. Joule's apparatus, schematically shown in Figure 13-2, consists of a water-filled vessel containing a rotating axis with several stirring paddles attached to it. The water in the vessel was prevented from rotating along with the paddles by special vanes attached to the walls of the vessel. The

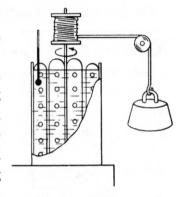

Figure 13-2. Joule's apparatus.

whole system looked very much like the modern gearless automatic transmission of an automobile and provided a perfect stage for the play of internal friction forces. The axis with the paddles was driven by a weight suspended across a pulley, and thus the work done by the descending weight was transformed into friction heat communicated to the water. Knowing the amount of water in the vessel, Joule could measure the rise of its temperature and calculate the total amount of heat produced; the driving weight and the distance of its descent gave the total amount of mechanical work done. Repeating this experiment many times and under different conditions, Joule established that there is a direct proportionality between these two quantities and that "the work done by the weight of 1 lb through 772 ft at Manchester will, if spent in producing heat by friction in water, raise the temperature of 1 lb of water 1°F." In metric units, it means that 1 cal of heat is the equivalent of $4.18 \times 10^7$ ergs of work.

Joule's work confirmed the basic idea that was in the air at the time—namely, that **heat is energy in the same**

**sense as mechanical energy is, and, while one form of energy can be transformed into another, the sum of the two always remains constant.** This law represents one of the basic pillars of the entire system of physics.

## Heat Conduction

If we take a long iron rod and heat one end of it, by means of a lighted candle, let us say, heat will propagate along the rod and gradually raise the temperature at points more and more distant from the heated end (Figure 13-3). If the rod is perfectly insulated by, say, a layer of asbestos, it will finally acquire the temperature of the flame throughout its entire length, and the flux of heat will stop. If the rod is permitted to lose heat through its surface to the surrounding air, however, there will be established a certain state of equilibrium with the temperature gradually dropping along the rod. If, instead of an iron rod we take a glass rod, the sequence of events will be the same except that the establishment of the final temperature distribution will take a considerably longer time; thus we say that glass is a poorer heat conductor than iron. The basic law of heat conduction states that **the rate of heat flow—that is, the amount of heat passing through a unit of cross-section per unit of time—is proportional to the gradient of the temperature**, and we can define the heat conductivity of different materials as *the number of calories passing through $1cm^2$ of cross-section per sec, if the temperature drops by $1°C/cm$.* The heat conductivities of several familiar materials are shown in Table 13-1, and we notice right away that metals are, in general, much better heat conductors than nonmetals. There is a very good reason for this, as we will learn later when we discuss the internal structure of matter and the hidden mechanism of heat conductivity.

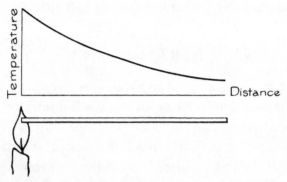

Figure 13-3. The relationship between distance and temperature in a heated rod.

TABLE 13-1 HEAT CONDUCTIVITY OF DIFFERENT MATERIALS EXPRESSED IN CAL/SEC FLOWING THROUGH 1 CM², WHEN THE TEMPERATURE GRADIENT IS 1°C/CM

| Material | Heat Conductivity (at 18°C) |
| --- | --- |
| Silver | 0.97 |
| Copper | 0.92 |
| Aluminum | 0.48 |
| Iron (cast) | 0.11 |
| Lead | 0.08 |
| Mercury | 0.016 |
| Glass | 0.0025 |
| Brick | 0.0015 |
| Water | 0.0013 |
| Wood | 0.0003 |
| Asbestos | 0.0002 |
| Cotton-wool | 0.00004 |
| Air | 0.00006 |

The heat conductivity of various materials plays an important role in all kinds of heat insulation. Since cotton-wool and similar materials present forty times more resistance to the flow of heat than ordinary brick, we can clearly see the advantages of their use for insulating homes. And, since the escape of heat is proportional to the surface of an object, to conserve heat it is advantageous to build houses as compact (as close to a spherical shape) as possible—hence the difference in construction styles in southern California and northern Canada. Following the same principle, many animals roll up almost into a ball when it is cold, and stretch out when it is too warm.

Another important point in heat conductivity pertains to the size of the building to be heated. If we compare a modern apartment house with a log cabin, we will see at once that, per unit space, the heating of the former is much more economical. In fact, if a large building has the same geometrical shape as a small one, the total number of rooms (each of which is to be heated) increases as the cube of linear dimensions, whereas the surface through which the heat escapes increases only as the square. Thus, in larger buildings, the amount of heat needed per room to maintain a comfortable temperature is considerably lower than that needed for smaller buildings. The same applies to animals of various sizes that maintain their body temperature by chemical reactions between the food they assimilate and the air they inhale (metabolism). In Figure 13-4 the metabolic rates of different animals are plotted with respect to their weight. We see that hummingbirds, which have a very unfavorable surface-to-volume ratio, have to metabolize at a terrific rate—which is, by the way,

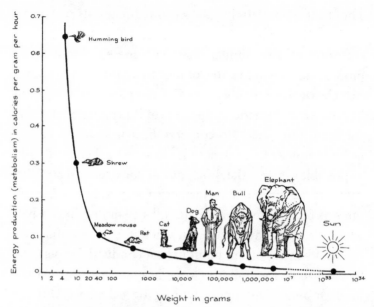

*Figure 13-4. The relationship between weight and heat production in various animals ranging in size from the hummingbird to the elephant. This relationship is extended to the right on the graph to include the Sun's interior.*

just about the same as the power production per unit weight in a modern helicopter. On the other hand, large animals can be very economical in their internal heating systems. In fact, if an elephant were suddenly to metabolize at the same rate as a hummingbird, it would soon be a roasted elephant, since the temperature of its body would rise to that usually encountered in kitchen ovens. In this connection it is interesting to note that the rate of "metabolism" inside the Sun (the dot on the far right of the curve in Figure 13-4) is extremely low—only a fraction of a percent of that in the human body. If the heating unit in an electric coffeepot produced heat at the same rate as it is produced in the interior of the Sun, the pot would take months to make water boil (assuming of course, that the coffee pot were perfectly insulated against heat losses). Only the extremely low surface-to-volume ratio of the Sun enables its meager rate of energy production to keep its body and surface much hotter than that of a roasted elephant!

## Heat Convection

In the case of poor heat conductors, the propagation of heat into the heated body is very slow, and in the case of water, for example, it would take hours to heat the water in the teakettle standing on the fire if there were no other heat-carrying processes. In fluids, the propagation of heat is considerably accelerated by the process of convection, which has its basis in the fact that heated bodies increase their volume and hence decrease their density. In our teakettle, the water near the bottom is heated by immediate contact with the hot metal, becomes lighter than the rest of the water in the kettle, and floats up, its place being taken by the cooler water from the upper layers. These convection currents carry the heat up "bodily," and they mix the water in the kettle so that the tea is ready in practically no time. A similar phenomenon takes place in our atmosphere when, on a hot summer day, the air heated by contact with the ground streams up to be replaced by cooler air masses from above. As the air rises to higher and cooler layers of the atmosphere, the water vapor in the air condenses into a multitude of tiny water droplets and forms the cumulus clouds so characteristic of hot summer days. Convection processes are also very important in the life of our Sun and stars. In them, the atomic energy produced in the central regions is carried surfaceward by streams of heated stellar gases.

Sometimes the notion of "convective" heat transfer gets mixed up with the notion of *heat conduction*. We have seen from Table 13-1, for example, that the heat conductivity of cotton-wool is about the same as that of air. Wool, fur, and other materials used to make warm clothes also have about the same degree of conductivity. But if the heat conductivity of air is the same as that of warm clothing materials, why then is a naked man less comfortable in cold weather than a man in a fur coat or under a thick woolen blanket? The reason is that the heat is removed from the skin of a naked man not by heat conduction in the air but by heat convection: the cold air warmed by contact with the skin rises and is replaced by more of the cold air. The role of warm clothing materials is to prevent this circulation and to keep the air from moving by trapping it between the numerous interwoven fibers of the materials. If we compress a woolen sweater or a mink coat under a hydraulic press, they will immediately lose their ability to keep us warm.

## Very High Temperatures

Whereas, as we have seen, there exists an absolute zero temperature (−273°C), there is no upper temperature limit. The filament of an ordinary electric bulb reaches a temperature of 2,000°C, but modern high-temperature engineering can produce temperatures up to about

15,000°C by using so-called "plasma-jet" torches. The surface temperature of the Sun is about 6,000°C, while the interior temperature is 20 million degrees.* In the core of an exploding A-bomb (uranium fission), the temperature is about 50 million degrees, while in an H-bomb (deuterium fusion) it is much higher.

**TABLE 13-2 MELTING AND BOILING POINTS OF METALS**

| MATERIAL | MELTING POINT (°C) | BOILING POINT (°C) |
|---|---|---|
| Tin | +232 | +2260 |
| Lead | +327 | +1620 |
| Aluminum | +660 | +1800 |
| Copper | +1083 | +2300 |
| Iron | +1535 | +3000 |
| Platinum | +1773 | +4300 |
| Tungsten | +3370 | +5900 |

|  | melting |  | boiling |  |
|---|---|---|---|---|
| Solid | ⟶ | Liquid | ⟶ | Gas |

Looking at Table 13-2, which gives the temperatures at which various materials melt (turning into liquid) and boil (turning into gas), we find that at extreme temperatures of heat, *all* chemical elements turn to gas.

## Very Low Temperatures

To produce very low temperatures, we use cryogenic equipment, which is similar to the mechanism of a refrigerator or a room air conditioner and is based on the

---

*When we speak about such high temperatures, it really does not matter whether we express them in °C or °K, which differ only by 273°.

fact that compressed air escaping through a small opening gets cooler when it expands into a larger volume. You can verify this fact by blowing air out of your lungs, first with your mouth wide open, and then with your lips forming a small hole. In the first case, the palm of your hand placed in front of your mouth will feel quite warm; in the second case, it will feel much cooler. The principle of a cryogenic apparatus is shown in Figure 13-5. The

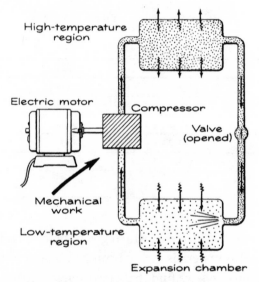

*Figure 13-5. The principle of cryogenics. The gas, heated by compression in the upper chamber, loses heat to the surroundings, while the gas that is cooled by expansion in the lower chamber absorbs heat from the surroundings. (Direction of heat flow is shown by wavy-tailed arrows.)*

apparatus consists of an electrically driven "compressor" that pumps the gas from the "expansion chamber" up into the "compression chamber" when the valve on the right connecting the two chambers is closed. The compressed gas is heated to above room temperature, and the excess heat escapes into the surroundings, be it into the kitchen in which the refrigerator stands or into the air outside the window in which the air conditioner is installed. When the valve separating the compression chamber and the expansion chamber is opened, the gas expands again and "sucks in" heat from the surroundings. Then air from the expander is brought back into the compressor and the process is repeated. Thus, the sequence of successive compressions and expansions serves as a "thermal pump" that takes the heat away from the expander and "pumps" it into the compressor. The refrigerator, of course, warms up the kitchen while it cools the food inside it, and the air conditioner warms up the outside air while it cools the room—although nobody seems to care much, unless the machines break down.

It is possible to build complex cryogenic equipment capable of reducing the temperature of air so much that it can first liquefy it and then freeze it into a solid block. In the same way, all other gases, such as hydrogen and helium, can be liquefied and frozen, but they require much lower temperatures. Table 13-3 gives the temperatures of the liquefaction and freezing points of several common gases.

### TABLE 13-3 LIQUEFYING AND FREEZING POINTS OF GASES

| Substance | Liquefying Point (°C) | Freezing Point (°C) |
|-----------|----------------------|---------------------|
| Chlorine | −35 | −102 |
| Oxygen | −183 | −218 |
| Nitrogen | −196 | −210 |
| Hydrogen | −253 | −259 |
| Helium | −269 | −272 |

## Turning Heat into Motion

We know that any amount of mechanical energy can be completely transformed into heat and, thus, that all the kinetic energy of a heavily loaded, fast-moving train is transformed into heat by the friction in the brakes when the train stops. But is the process reversible? Can the entire amount of heat contained, let us say, in a pot of boiling water be transformed into mechanical energy? We know for sure that steam engines transform heat into mechanical energy, but if we look at the problem closer we will find that they always transform only a *part* of it. What happens to the other part? Well, as any mechanical engineer will tell you, it is delivered into the "cooler" which receives the steam after it has done its work in the cylinders. Hot steam produced in the boiler is sent into the cylinder (by opening the "incoming" valve), pushes the piston, and thus transforms a part of its thermal energy into mechanical work. The piston then moves and pushes the steam that has been used through the "outgoing" valve into the cooler where it condenses back into water.

"Coolers" are always used in steamship engines; since sea water is no good for running steam engines (too much salt deposits on the boiler walls), the water collected in the cooler is again transferred to the boiler. In railroad locomotives, the role of the "cooler" is played by the surrounding air into which the steam is released from the cylinders (the air acts as a "cooler," of course, only as long as it is cooler than the boiler). Summing up, we see that the principle of the operation of a steam engine can be formulated in the following way: **The heat from the hot region (the boiler) streams down into the colder region (the cooler) and, on the way, a certain part of it is transformed into mechanical energy.** If $Q_1$ is the amount of heat coming from the boiler, and $Q_2$ the amount received by the cooler, the amount of mechanical energy produced is given by:

$$E = (Q_1 - Q_2) \times (\text{mechanical equivalent of heat})$$

From the point of view of economy, of course, it would be best to reduce $Q_2$ to zero, throw away the cooler, and have all the heat contained in the hot steam turned into mechanical energy. But this is impossible. To understand why, let us consider the following problem, which at first sight has nothing to do with heat or steam engines. Suppose there is a house on a hill, and a creek running swiftly through a ravine a dozen feet below it (Figure 13-6). Can we manage it so that the creek by its own power will supply the household's water needs? The answer is: yes. If we build a dam $A$ and install a waterwheel $B$ that will produce a certain amount of power, the waterwheel can operate a pump $C$ that will pump a certain amount of water up the hill and into the house. Very simple indeed! But, if the owners of the house become too ambitious and try to get all the water carried by the creek up the hill, they will be heading for trouble. The amount of water they are getting is being pushed 12 ft uphill by the rest of the water dropping 3 ft in the waterwheel. If all the water in the creek were brought up to the house, there would be no water left to drive the wheel and to operate the pump! The best we

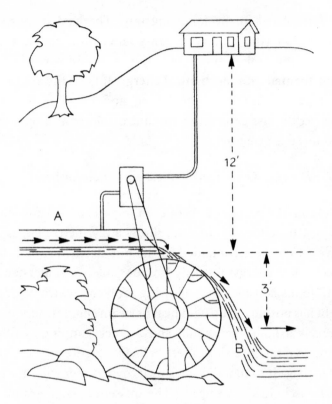

Figure 13-6. Pumping water uphill by means of a water-wheel.

can do is to arrange things in such a way that the potential energy liberated by the water operating the wheel is the same as the potential energy necessary to raise the water to the house. If $X$ is the fraction of the total water supply of the creek that can be brought up to the house, we have:

$$X \times 12 = (1 - X) \times 3$$

so that:

$$X = \frac{3}{12 + 3} = \frac{1}{5}$$

Thus, at best, we can get one-fifth of the water of the creek "self-propelled" to the house, but any demand beyond that would contradict the laws of physics.

The simple situation with the steam engine is quite similar, and it was shown by a French engineer, Sadi Carnot (1796-1832), that **the largest fraction of original heat "descending" from the temperature $T_1$ to the temperature T2 that can be turned into mechanical energy is, at best, equal to the ratio of the temperature difference to the higher temperature $(T_1 - T_2)/T_1$, where** the temperatures are counted from absolute zero. If, for example a locomotive engine operates between a boiler temperature of 100°C and an outside temperature of say, 40°C (373 and 313 in the absolute scale), the maximum fraction of heat from the boiler that can be transformed into mechanical work is:

$$\frac{373-313}{373} = \frac{60}{373} = 0.16, \text{ or about 16 percent}$$

The simple reciprocating steam engine we have discussed, in which the "to and fro" motion of a piston is transformed into the rotational motion of a crankshaft, was invented during the second half of the eighteenth century by a Britisher, James Watt, and it has been used extensively ever since for stationary power production as well as for the propulsion of steamships and locomotives. More recently, reciprocating steam engines have yielded, in many instances, to *steam turbines*, which have the advantage of continuity of motion and the absence of valves. The scheme of a steam turbine is shown in Figure 13-7. It consists of a fixed drum, *A*, on the left, containing tilted metal paddles, fixed in place and arranged in such a way that the high-pressure steam that passes through the drum for the boiler whirls around them in a counterclockwise direction. Another similar, but movable, drum *B* is placed right in front of drum *A* and contains paddles that are tilted opposite of those in *A*. The impulse of the hot vapor streaming from drum *A* causes drum *B* to rotate in the direction indicated by the arrows.

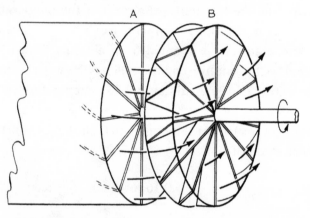

*Figure 13-7. The scheme of a steam turbine. Steam turbines operate by using hot water vapor, while gas turbines use the hot gases that result from the burning of fuel.*

## Mysterious Entropy

If we could transform 100 percent of a given amount of heat into mechanical energy, we would be almost as well-to-do as we were with the "perpetual motion" machines described in the chapter on mechanical energy. Instead of "producing energy from nothing," we could turn the heat content of any surrounding medium into mechanical energy. An ocean liner could pump in sea water, extract the heat energy from it to drive its propellers, and throw overboard the resulting block of ice. An airplane could take in air, turns its heat content into kinetic energy, and throw the ice-cold jet out through the nozzle in the rear. In fact, since the air, the water, and the ground are heated well above the temperature of absolute zero, these "perpetual motion machines of the second kind" would be just as effective as the "perpetual motion machines of the first kind." But, as we have seen before, such machines are also impossible. We cannot use the heat content of our surroundings to produce mechanical work any more than we can use the water in the oceans to run hydropower installations. The potential energy of the water in the oceans is useless because there is no lower water level to which this water can flow; the heat content of our surroundings is useless because there is no lower temperature region to which this heat can flow.

Summarizing the above fact, we may say that **the natural direction of heat flow is from hot regions to cold regions and the natural direction of energy transformation is from mechanical energy into heat energy**.

In the "natural" direction, both processes can proceed by themselves 100 percent, but if either of them is to run in the "unnatural" direction, there must always be an accompanying process running in the "natural" direction which is sufficient to compensate for the "unnatural" behavior of the first one. Thus, it is "unnatural" for heat to escape from the cool interior of a refrigerator into the warm kitchen air, but this process can take place because it is compensated for and even overcompensated for by the "natural" transformation of the electric energy driving the refrigerator's motor into heat. It is "unnatural" for the heat of the steam in a locomotive's boiler to go over into mechanical energy and drive the wheels, but, here again, the "unnaturality" of the process is overcompensated for by the "naturality" of the flow of a part of the heat from the hot boiler to the cool air outside.

In *thermodynamics*—i.e., the study of the relation between heat and mechanical motion—the degree of "naturality" of heat transformation is called *entropy*. We say that the entropy increases when the process goes in the "natural" direction and decreases when it goes in the opposite direction. When a hot object cools down upon being thrown into cold water or an automobile is stopped by application of its brakes, the total entropy of the system goes up. In the work of steam engines or refrigerators, the entropy of the working parts (cylinder and piston or the cooling unit) of the machine goes down, but it is compensated for (or overcompensated for) by an increase of entropy elsewhere. **But in the grand total, the entropy of the entire system cannot decrease and it either remains constant or, in most cases, goes up.** If this were not so, engineers would be able to construct the fabulous perpetual motion machines of the second kind, described in the beginning of this section, and we would have an unlimited supply of free energy for industrial and other uses. Technically, the change of entropy of a given body is measured by the amount of heat it gains or loses (a negative sign in the second case) divided by its (absolute) temperature. In the earlier example of the steam engine that operated between the temperatures $T_1$ and $T_2$, the change of entropy of the boiler when the amount of heat

$Q_1$ is taken from it is $-Q_1/T_1$, and the corresponding change in the cooler is $+Q_2/T_2$. Since the total change of entropy must be equal to or larger than zero, we easily arrive at the previously quoted Sadi Carnot's Law. In fact, we can write:

$$\text{Entropy change} = \frac{Q_1}{T_1} + \frac{Q_2}{T_2} \geq 0$$

so that:

$$\frac{Q_2}{T_2} \geq \frac{Q_1}{T_1} \, or \, \frac{Q_2}{Q_1} \geq \frac{T_2}{T_1}$$

The efficiency $\eta$ of the engine is given by:

$$\eta = \frac{Q_1 - Q_2}{Q_1} = 1 - \frac{Q_2}{Q_1} \leq 1 - \frac{T_2}{T_1} = \frac{T_1 - T_2}{T_1}$$

The reader must have noticed that both examples cited to point out the local decreases of entropy (a steam engine and a refrigerator) are manmade machines. Indeed, all, or almost all, processes in nature run in "natural" directions, with the entropy increasing more or less uniformly everywhere. The trick of producing local entropy decreases that are compensated for by an increase elsewhere is essentially the product of human ingenuity, the brainchild of clever engineers. The notable exception is that of living organisms, which operate on principles very similar to those used in manmade machines. But in building his machines, of course, man simply imposes on inorganic matter the same ingenious principles that operate in his own body.

## Thermodynamics of Life

At first sight it would seem that the Second Law of Thermodynamics fails in the case of living organisms, since the word "organism" itself implies a high degree of order and organization of the molecules forming it. Consider a plant growing from a seed inside a glass box containing soil at the bottom and abundantly supplied with fresh water and air, which brings in the carbon dioxide necessary for building new material for the growing plant. Although both water and carbon dioxide gas possess a very low degree of order, the plant nevertheless manages to organize H, O, and C atoms and to turn these simple substances into highly complex organic compounds such as sugars, proteins, etc. Don't we witness here a case where disorder goes spontaneously into order, with the entropy decreasing in contradiction to the basic rules of thermodynamics? The answer is: No. We have forgotten that, apart from water and a carbon dioxide supply (plus a small amount of some salts from the soil), the plant needs for its growth and development an abundance of sunshine. The Sun's rays that are absorbed by the green leaves of the plant bring in the energy necessary for building up complex organic molecules from the simple molecules of $H_2O$ and $CO_2$; but this is not the point. We are not worried about the first law of thermodynamics (i.e., conservation of energy), but about its second law, which prohibits any spontaneous decrease of entropy.

During the transformation of $H_2O$ and $CO_2$ into complex organic molecules, the entropy of the system decreases* and, according to the Second Law of Thermodynamics, we have to look for some related process that results in an equal or larger entropy increase. In fact, we find that, apart from supplying a growing plant with necessary amounts of energy, the Sun's rays also take care of the necessary entropy decrease. The point is that when solar radiation arrives at the surface of the Earth, it is strongly "diluted" in the sense that, whereas its *prevailing wave length* still corresponds to the temperature of the solar surface (i.e., 6,000°K), its *intensity* is not more than that of the radiation emitted by a room-heating radiator. A thermodynamical study of radiant energy indicates that such a dilution of the radiation without a reduction of the prevailing wave length leads to a high entropy *deficiency* or, as we sometimes put it, to the presence of "negative entropy." It is the inflow of this "negative entropy" that permits a plant to grow by organizing the $H_2O$ and $CO_2$ into more complex organic molecules. The processes of the growing of a plant under the action of the rays of the Sun and the subsequent burning of the material so produced in the fireplace can best be described by the following two symbolic equations:

$$\overset{\text{from soil}}{H_2O} + \overset{\text{from air}}{CO_2} + (\text{energy} + \overset{\text{from the sun}}{\text{entropy deficiency}}) \rightarrow \text{wood} + O_2$$

$$\text{wood} + \overset{\text{from air}}{O_2} \rightarrow (\text{energy} + \text{entropy excess}) + H_2O + CO_2$$

The energy, which *is to be conserved*, is absorbed in the first process and liberated in the second. The entropy, which *must always increase*, increases in both processes, since,

---

*The fact that a piece of wood burns spontaneously when set afire proves that (wood) + $O_2 \rightarrow H_2O + CO_2$ is the *natural* direction of the process.

indeed, the "disappearance of a deficiency" is equivalent to the "appearance of an excess."

The high organization of molecules (negative entropy), obtained by plants from the Sun's rays is then passed, along with accumulated energy, to animals, which in this sense are entirely parasitic beings, at least from the point of view of the plants.

# CHAPTER 14

# Electromagnetism

## *Atmospheric Electricity*

Electric phenomena have been known to man from time immemorial. Our earliest ancestors were horrified by lightning striking down from thunderclouds and ascribed the flash to the fury of the gods. But, even in the absence of a thunderstorm, strong electric tensions exist above the surface of the Earth, and we can produce miniature artificial lightning any time we want. To do this, we must collect electricity from the layers of air high above the ground, and discharge it into the ground through a wire. This can easily be done by placing an "electric collector" at the end of a nonconductive bar protruding from the roof of some tall building and running an insulated wire down from it to the ground. An electric collector is simply a metal plate covered with a thin layer of some radioactive material (the luminous paint used for watch dials would do) which makes the surrounding air electrically conductive and thus permits electric charges to flow toward it. If we wait for a while until enough electric charge is collected by such a device, we can produce an impressive spark by touching the end of the wire with our finger. Long ago it was found that electric tension in the air above the ground is about 100 volts (v) for each meter of height, so that the *electric tension* between the head and the feet of a standing man is more than the voltage in an electric plug in the wall. Why then do we pay our electric bill instead of connecting the outlets of electric appliances to the roof and to the basement floors of our homes? The answer is that *electric tension* alone is not enough to run electric bulbs or motors; we must also have sufficiently large *amounts of electricity* flowing under this tension.

A simple hydrodynamical example will clarify the situation. Consider a hydroelectric installation, such as that at Niagara Falls, in which 5,000 tons of water come down every second from a height of about 50 meters. As we have seen, the mechanical power of this installation is the product of the two given quantities and equals 250 million kg-m/sec. But, to produce that power we need both the large amount of water and the large height of fall. A glass of water per second will not produce much power even if dropped from the top of the Empire State Building, and all the water of the Mississippi River would be useless for producing power if this water fell only a few centimeters. Similarly, high electric tension cannot produce much power if there is not enough electricity, because *electric power is the product of the electric tension (or potential difference) and the strength of electric current*. Thus, any attempt to use the 40,000 volts of electric tension that exist in the air between the base and the top of the Empire State Building for producing power is just as useless as building a hydroelectric installation at a creek that falls from a height many times that of Niagara Falls but carries only a few barrels of water per minute.

## Triboelectricity

In the early experiments with electricity carried out by William Gilbert (1544-1603), a personal physician of Queen Elizabeth I, electric charges were produced mainly by "rubbing a galosh against a fur coat" (Figure 14-1) or a glass stick against a silk handkerchief. When a lady with a fur coat tries to get into a car with plastic seat covers or a gentleman wearing rubber-soled shoes walks across carpet, sparks may fly when they touch the handle of the car door, or the radiator in the room (this happens only when the air is dry, however, since humidity increases the electric conductivity of air and thus helps to neutralize electric charges). Electricity produced in this way is known as *triboelectricity*, and it served to establish the first laws of electric interactions.

*Figure 14-1. An elementary experiment with triboelectricity. From* Physics for Fools, *published in St. Petersburg, Russia, in 1908.*

If we suspend side by side two light metallic spheres and touch them both with a hard rubber stick rubbed against a piece of fur, we will find that the two spheres repel each other (Figure 14-2a). The same happens if we touch both spheres with a piece of fur against which the stick was rubbed (Figure 14-2b). However, if one sphere is touched by the hard rubber stick and the other by the fur, the two spheres will attract each other (Figure 14-2c). On the basis of these elementary experiments, Gilbert concluded that there are two kinds of electricity and that **electric charges of the same kind repel each other while those of the opposite kind attract each other.** He called electric charges produced by friction on the fur *positive*, and those produced on the rubber *negative*. Studying in more detail the interactions between electric charges, the French physicist, C. A. Coulomb (1736-1806), found that **the force of attraction or repulsion between two electrically charged bodies varies in direct proportion to the product of their charges\* and in inverse proportion to the square of the distance between them.** This so-called *Coulomb's Law* formed the cornerstone for all further studies in the field of electricity. Following the general method of defining the units for new physical quantities, *we can define the unit of electric charge as the amount of electricity which acts with a force of 1 dyne on an equal amount placed 1 cm away.* This unit, known as an *electrostatic unit* of electricity, is, however, too small for practical purposes, and in electrical

---

\*Coulomb could experiment with half-unit, one-third-unit, and smaller integral fractions of charges by taking a charged sphere and bringing it in contact with one or more uncharged ones of equal size.

engineering a much larger unit known as a *coulomb* is used, which is equal to 3 billion ($3 \times 10^9$) electrostatic units. The physical dimension of an electrostatic unit is:

$$|e| = \frac{|gm|^{\frac{1}{2}} \cdot |cm|^{\frac{3}{2}}}{|sec|}$$

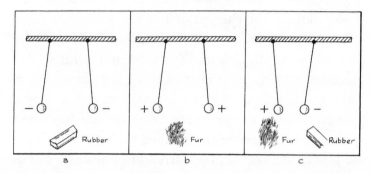

Figure 14-2. *The repulsion and attraction between electrically charged bodies.*

## *Electric Current*

If we connect a conductor charged with, let us say, positive electricity (fur's kind) with the ground by a metallic wire (Figure 14-3*a*), the conductor will be discharged, and for a split second an electric current will flow through the wire. The duration of such an electric discharge is, however, too short for convenient study, and it is desirable to have an arrangement that provides us with a steady current. This became possible after the discovery that steady electric current can be produced by the so-called *electric cell*, invented by an Italian physicist, A. Volta (1745-1827). It consists of two plates made of two different substances, such as carbon and zinc, placed in some acid solution such as sulfuric acid (Figure 14-3b). During the early studies of electricity, it was believed that electric charges were caused by a hypothetical "electrical fluid" and that the positive and negative charges correspond to the excess and deficiency of that fluid in the charged object. The phenomenon of triboelectricity was interpreted as the transfer of a certain amount of that fluid from the rubber to the fur in the process of rubbing. Since in an electric cell made of carbon and zinc plates, the electric charge appearing on the carbon plate is similar to that produced on the fur and the charge appearing on the zinc plate is similar to that produced on the rubber, the cell was considered to produce an excess of electric fluid on carbon and a deficiency of that fluid on zinc. Thus, the current was considered as the flow of electricity from the carbon electrode (known as the *anode*) to the zinc electrode (known as the *cathode*).

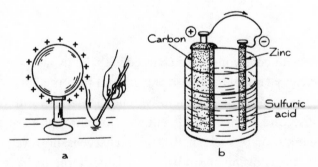

Figure 14-3. (a) An instantaneous electric current from a charged conductor; (b) a continuous electric current from a chemical battery of a very simple type.

As we shall see later in the book, the actual situation that exists when an electric current flows through a metallic wire is exactly the opposite. Electricity is actually carried by a flow of tiny negatively charged particles, the electrons, which make their way through the lattice of atoms forming metallic bodies, and although according to conventional terminology the current flows from anode to cathode, the actual motion of the charged particles takes place in the opposite direction. It would, however, be pointless to change the terminology because of our better knowledge of electric phenomena. In fact, when an electric current passes through a liquid or a gas, it is carried by both positively and negatively charged particles moving in opposite directions.

We *can define the unit of electric current as a flow of one unit of electricity per unit of time.* Thus, the CGS *unit of electric current* is the current that carries 1 electrostatic unit of charge/sec, while the practical unit, or *ampere (amp),* is a current carrying 1 coulomb/sec. Earlier in this chapter we introduced the notion of *electric tension,* or the *difference of electric potentials* that makes the current flow, and defined the *power* of an electric installation as the product of the current flowing through it and the electric tension. The practical unit of electric tension is the *volt (v), defined as an electric tension which, in conjunction with a current of 1 amp, produces 1 watt (w) of power.* Thus a 100-w bulb in a table lamp operating under a tension of 110 v takes a current slightly less than 1 amp.

## Terrestrial Magnetism

The fact that certain natural iron ores, known as "lodestones," when suspended on a string, assume a definite position with one end pointing roughly to the north pole and another to the south pole, was known to the ancient Chinese. The magnetic compass developed on that principle is of immense value for finding one's way, both for the ships sailing across the oceans of the world and for boy scouts lost in the woods. The magnetic field of the Earth that orients the needle of the compass manifests itself in many other ways; for example, it deflects toward the poles the beams of electrically charged particles which come to us from the Sun, thus producing the magnificent phenomena of *aurora borealis* (polar lights).

We can use the magnetic field of the Earth to "magnetize" steel rods by holding them in the direction of the magnetic field of the Earth and hitting them repeatedly with a hammer. The violent impacts shake the tiny particles constituting the internal structure of metal and orient them, at least partially, in the direction of the field. As a matter of fact, all steel objects possess a certain small degree of magnetization induced by the terrestrial magnetic field, and during the war much effort was spent to "demagnetize" warships and transports so they would not trigger the magnetic mines laid by the enemy.

If we bring close together two magnetized steel rods, we will find that *the "homologous" ends, i.e. the ends that pointed the same way during the magnetization process, repel each other and that if one of the rods is turned around, the ends of the rods will attract one another*. Since, according to accepted terminology, the end of a compass needle or any other magnet that points north is called its *north pole*, we conclude that the magnetic pole of the Earth located near its geographical north pole is actually a magnetic south pole, and vice versa. But, again, as in the case of positive and negative electricity, it would be too much trouble to change terminology.

At this time, it is important for us to remember the fact that, unlike positive and negative charges, *magnetic poles always occur in pairs* and that it is impossible to cut a north or south pole from a magnet and carry it away, for if we cut a magnet into two pieces we will get two smaller magnets, since a new pair of poles will originate at the broken ends. The celebrated British physicist, P. A. M. Dirac, who predicted on the basis of purely theoretical considerations the existence of the so-called "positive electrons," had a theory according to which single magnetic poles should be found to exist in nature. However, the existence of these so-called "magnetic monopoles" has not been experimentally verified.

The attractive and repulsive forces magnetic poles obey a law similar to that of electric charges: *the forces between the magnetic poles are directly proportional to the product of the strengths of the poles\* and inversely proportional to the square of the distance between them*. We can also define a unit of magnetism (*or the strength of a magnetic pole*) as *the amount of magnetism that repels with a force of 1 dyne an equal amount placed 1 cm away*.

---

\*One can double or triple the strength of the magnetic pole used in such experiments by fastening together several equal small bar magnets.

## Electromagnetic Interaction

The possibility of having a continuous flow of electricity led to the study of various interactions between electric currents and magnets. (Stationary electric charges and magnets do not interact at all.) There are several basic laws governing these interactions, all of them discovered early in the nineteenth century. In the year 1820, a Danish physicist, H.C. Oersted (1770-1851), noticed that an electric current flowing through a wire deflects a magnetic needle placed in its neighborhood in such a way that the needle assumes a position perpendicular to the plane passing through the wire and through the center of the needle (Figure 14-4). Oersted's discovery was followed up by two French physicists, J. B. Biot (1774-1862), and F. Savart (1791-1841), who amended it by the statement that the forces attempting to orient the needle in this direction are directly proportional to the strength of the current and inversely proportional to the distance of the needle from the wire. This law provided a simple method for detecting and measuring electric currents by means of a galvanometer, the scheme of which is shown in Figure 14-5. It consists of a coiled wire that can carry an electric current, and a magnetic needle placed in the center of the coil in such a way that its normal position (as determined by the suspension wire) is in the plane of the coil. When a current flows through the coil, the forces of electromagnetic interaction attempt to turn the needle into a position perpendicular to the plane of the coil, against the resistance of the twisted suspension wire. Thus, the stronger the current, the larger is the angle through which the needle will be turned away from its normal position, and by measuring that angle we can find how strong the current is.

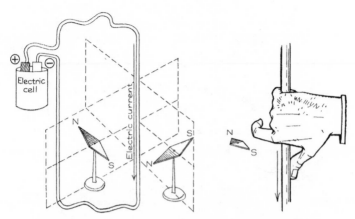

Figure 14-4. *The orientation of magnets (compass needles) in the neighborhood of an electric current. The direction in which the north pole of the needle will point can be found by the following rule: Grab the wire with your right hand so that the thumb is in the direction of the current; the index finger will then indicate the direction of the needle's north pole.*

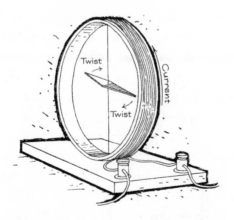

Figure 14-5. *The principle of the galvanometer. The current in the coil deflects the magnetic needle against the resistance of the vertical suspension wire.*

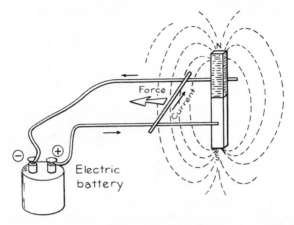

*Figure 14-6. A movable wire carrying electric current (black arrows) experiences a mechanical force (white arrow) if placed in the neighborhood of a magnet. If the direction of the current is reversed, or if the magnet is turned upside down, the direction of the force will be reversed too.*

Suppose now that instead of a movable magnet and a fixed wire carrying an electric current, we have the reverse situation of a fixed magnet and a movable wire (Figure 14-6). Since the magnet cannot move but the wire can, electromagnetic interaction will result in the motion of the wire as indicated in the figure. If the direction of current is reversed or if the magnet is turned upside down, the direction of the force acting on the wire will also be reversed. Experiments of this kind proved that the force acting on a current-carrying wire placed in the field of a magnet is directed perpendicular to the length of the wire as well as to the direction of the magnetic field (i.e., the direction that would be assumed by a compass needle in that spot) and is proportional to the strength of the current. The interaction between currents and magnets constitutes the principle of the electric motors. In one type of motor the wire is coiled on an armature which can rotate in a strong magnetic field. The coiled wire is arranged in such a way that the forces of electromagnetic interaction between the current it carries and the magnetic field tend to turn it by a certain angle. When this turn is accomplished, however, the direction of the current is automatically reversed, and the motion continues indefinitely. The situation is similar to that of a donkey who runs after a carrot suspended a few feet in front of him from a stick tied to his neck.

Now we come to a very important modification of the experiment just described. Suppose we change the arrangement from that shown in Figure 14-6 to that shown in Figure 14-7 by substituting a galvanometer for the source of the electric current, and, instead of letting the wire move at will, we move it by hand toward or away from the magnet. What will happen? There is in physical science a very general principle known as the *Principle of Le Chatelier*, which, in a way, represents the generalization of the mechanical principle of inertia. It state that **whenever we try to impose changes on the "status quo" of a system, the "nature of things" will do its best to oppose our action**. To start with, the movable wire was resting on its supports, the galvanometer was showing no current, and everything was nice and quiet. But once we disturb the system by moving the wire toward or away from the magnet, something must happen to prevent us from doing so. What happens is that, as the result of our act, a current is induced in the moving wire, and the direction of that current is such that the current-magnet interaction creates a force opposing the motion of the wire. The fact that **electric currents** are **always produced (or induced) in any electric conductor moving through a magnetic field** constitutes the principle of all kinds of dynamos and generators, in which electric currents are originated by rotating in strong magnetic fields the wire coils attached to the armatures. Just as the experiments shown in Figures 14-6 and 14-7 are mutually complementary, electric motors and electric generators are the counterparts of one another. If we rotate the armature, a current will be induced in the wire; if we send a current through the wires in the armature, the armature will rotate.

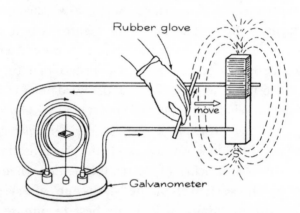

*Figure 14-7. If the wire is forcibly moved in the field of a magnet, an electric current will be induced opposing that motion.*

So far we have been talking about the interaction between currents and magnets. What about the interaction between two currents? This case was first studied by the French physicist, A. M. Ampère (1775-1836), who showed that **two wires carrying electric currents which flow in the same direction are attracted to one another, whereas**

**in the case of currents flowing in opposite directions, the attraction changes to repulsion** (Figure 14-8). This law of the interaction between currents can be logically derived from the previously stated laws of electromagnetic interactions. Electric current flowing through one of the wires produces a magnetic field which acts on the current in the second wire, pulling or pushing it away from the first one.

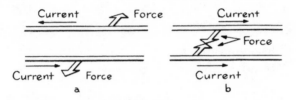

*Figure 14-8. The mechanical forces acting on two wires carrying electric currents.*

Ampère also found that, **in the case of two parallel wires, a current sent through one of them produces a current of opposite direction in the other** (Figure 14-9). This *induced current* exists, however, only for the short time during which the current in the first wire is building up from zero to its maximum value. But if we cut off the current in the first wire, another current (in the same direction as that in the first wire) will be induced in the second wire when the first current is dropping down to zero. This law, too, can be derived from what was said above about the interactions of currents and magnets. Indeed, when the current is growing in the first wire, the magnetic field around it becomes stronger and stronger. Thus, the second wire finds itself in a rapidly increasing magnetic field which must have the same effect as if it were moving toward a permanent magnet. The currents induced in one wire by a change of the current in the other play an important role in many practical applications, and we will encounter this phenomenon again later in this chapter when we discuss oscillating circuits.

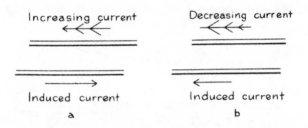

Figure 14-9. The induced current produced in one wire by increasing and decreasing the current in another wire.

## The Nature of the Electromagnetic Field

In considering the mechanical interactions between material bodies, we are accustomed to the fact that such interactions require immediate bodily contact. If we want to move an object, we have to touch it with our hand, or else have a stick to push it or a rope to pull it. On the basis of such views, the famous British physicist, Michael Faraday (1791-1867), to whom science is obligated for many important discoveries in the field of electricity, liked to imagine that what one usually calls "empty space" is actually filled by some peculiar substance, "world ether," which is responsible for all electric and magnetic interactions. According to Faraday's conception, the placing of an electric charge on a copper sphere or the magnetization of an iron bar results in certain deformations in the surrounding "world ether" that lead to repulsive or attractive forces between the material bodies. The stress and strain lines in this hypothetical "world ether" were supposed to coincide with the *lines of force* defined by the direction of electric or magnetic forces in the point of the surrounding space. In Figure 14-10a and *b*, we see the lines of force in the space around two unlike and two like electric charges. If, following Faraday, we visualize these patterns as being formed

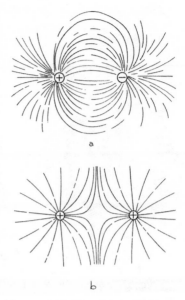

Figure 14-10. The electric field around (a) opposite and (b) identical electric charges.

by some elastic rubber-like bands, it is easy to get the impression that in the first case the stresses will lead to an attraction and in the second to repulsion. Somewhat more complicated patterns are shown in Figure 14-11. These photographs represent the lines of force surrounding the two magnets when they are oriented parallel and antiparallel to one another. Here again one may feel that the first case corresponds to repulsion and the second to attraction. Faraday's views were put into mathematical form by his disciple, James Clerk Maxwell (1831-1879), who proved that electromagnetic interaction can be represented by a set of equations that can be interpreted as describing the stresses and strains in some elastic medium.

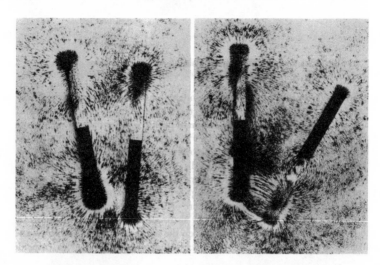

*Figure 14-11. The magnetic fields of two magnets in parallel and anti-parallel positions, shown by the orientation of iron filings. Photograph by R. L. Conklin, Hanover College, Ind.*

Although Maxwell's equations represent the basis of today's theory of electricity and magnetism, their interpretation has been radically changed. As will be described in more detail in Chapter 16, Einstein's Theory of Relativity rejected the notion of an all-penetrating "world ether" as a physical reality and returned to considering emptiness as the basic property of space. On the other hand, Einstein ascribed a physical reality to the *electromagnetic field*, and considered it to be some peculiar kind of material surrounding electrified and magnetized objects and responsible for the interactions between them. Instead of extending through entire space, however, this electromagnetic material exists only where electric and magnetic forces are present and is absent in really empty space. Thus, the field around an electrically charged conductor or a magnet should be considered as a jelly-like material surrounding them in the form of a local cloud rather than as local deformations in a jelly-like medium which fills up all of space (see Figure 16-6).

## *Electromagnetic Oscillations*

If we take two spherical electric conductors and give their surfaces opposite electrical charges, we will have to do work in some way to pull electrons off the surface we charge positively and to force excess electrons onto the surface we charge negatively. If we look to see where this work has gone, we will find that it is stored in the *electric field* which now exists between these two conductors. Suppose we connect these conductors by a piece of wire (Figure 14-12*a*). The opposite charges on the spheres will begin to be neutralized by the flow of electric current from one to the other, and as the charge on each sphere becomes less, the electric field between them will also decrease. We may ask, now: What happens to the energy stored in the electric field as the field becomes weaker? The answer is that we can find this missing energy stored in the *magnetic field* that has been created by the current in the connecting wire (Figure 14-12*b*). In Figure 14-12*c* we see the situation when the charges have been completely neutralized; the electric field has vanished, and all of its energy is now in the magnetic field. There is no difference in charge to keep the current flowing, and if this were all of the story, the current would stop. But Le Chatelier's Principle now goes to work to maintain the status quo—i.e., to prevent the

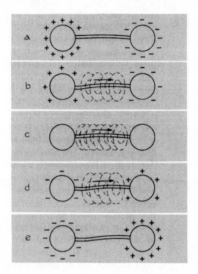

Figure 14-12. Electromagnetic oscillations between two charged conductors.

current from stopping. Hence the magnetic field delivers its stored energy back into the wire to keep the current flowing (Figure 14-12d), and the charge begins to build up on our spheres in the opposite direction. Finally, when the magnetic field has been reduced to zero, the current stops, and we have (Figure 14-12e) the same situation we had in the beginning (Figure 14-12a), except that the sign of the charges has been reversed. Now, of course, the whole cycle will repeat itself in the reverse direction, and again and again, and so on.

This oscillating electric current is analogous to the behavior of a pendulum. We start it by pulling the pendulum aside and giving it potential energy (we give the conductors the energy stored in the electric field). As it swings down, this potential energy is converted into kinetic energy (the energy of the electric field is now in the magnetic field). The pendulum swings over to the other side, and its kinetic energy is again converted to potential energy (the magnetic field has vanished, and its energy appears in the electric field again). And, just as the pendulum's period can be changed by changing its length, the period of electric oscillation in our system can be changed by changing the size of the two spheres or the distance between them.

The pendulum, of course, will not swing forever; it gradually loses its energy by friction. The electrons moving back and forth in our oscillating electrical system encounter resistance, too, in the wire through which they flow, and unless its energy were periodically replenished, it too would stop.

This analogy can be carried a step further. If we suspend our pendulum from a clothesline instead of a rigid support, some of the pendulum's energy will go to moving the clothesline back and forth and will be dissipated in waves traveling out along the line in both directions. Part of the energy of our oscillating electric circuit will in a somewhat similar way go into creating "electromagnetic waves" that radiate out into space. From the point of view described in the previous section, we can say that the "lumps" of jelly-like electromagnetic field material vibrating in the space surrounding the two spheres are torn away and travel freely into the space beyond. Here again, a propagating electromagnetic wave should be visualized as a vibrating lump of electromagnetic field material flying through empty space rather than as the propagation of an elastic deformation in some all-penetrating medium.

The existence of electromagnetic waves, which were predicted by Maxwell's theory, was proved in 1888 by the experiments of the German physicist, H. Hertz (1857-1894), and their practical importance was realized by an Italian engineer, G. Marconi (1874-1937), who established radio communication across the British channel in 1899 and, in 1901, across the Atlantic.

Electric circuits used in radio and TV transmitters today are just an improvement on the original scheme shown in Figure 14-12. First of all, in order to store a maximum amount of positive and negative electricity on two conductors, it is important to bring these two conductors as close to one another as possible. Thus, instead of two spheres which can be close to one another at only one point, it is preferable to use two metal plates located close together; such a pair of plates is known as an *electric condenser* (Figure 14-13). Also, in order to increase the amount of energy stored in the magnetic field, we use a longer wire and then coil it into what is known as a *solenoid*.

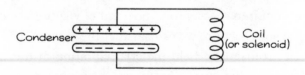

*Figure 14-13. An electromagnetic oscillator.*

# CHAPTER 15

# Light, Visible and Invisible

The study of light is one of the most important parts of physics since, indeed, most of our knowledge concerning the world around us is gained through seeing. We learn about the properties of giant stellar systems by means of light that travels for millions of years through empty space to deliver us its message. We learn about the properties of minute atoms through the light that is emitted by them and that carries in a hidden form important information concerning their internal structure. And, of course, most of the information that we get in our everyday life is also obtained through the medium of light.

## *Reflection of Light*

To demonstrate the basic laws of the propagation of light, we use thin light beams that can be formed by a screen with a small hole or slit in it. To make a light beam visible when we look at it from the side, we may blow tobacco smoke into it or arrange for it to graze obliquely along a white surface. The arrangement shown in Figure 15-1 is very useful for the study of the basic laws of the propagation of light. It consists of a vertical disc of frosted glass with the degrees of the angular scale plotted on it and a movable source that emits a thin light beam. At the center of the disc we can attach different devices such as mirrors, prisms, etc. that can also be rotated around the axis. If we use first a flat mirror, we will find that the direction of the reflected beam changes in respect to the mirror with the changing position of the light source. However, no matter how we place them, we will always find that the incident and the reflected beams of light form equal angles with the surface of the mirror.

While this rule holds for mirrors and, generally speaking, for all polished surfaces, light reflected from such surfaces as paper or frosted glass is scattered irregularly in all directions. This diffuse reflection of light is due to the fact that non-polished surfaces are coarse, being covered by a network of microscopic irregularities that have the same effect on the reflection of light as the surface of an unkept tennis court has on the direction of bounce of tennis balls.

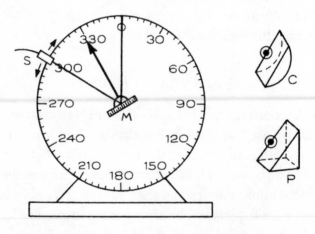

*Figure 15-1. A gadget for study of the reflection and refraction of light. (S) light source, (M) a mirror, (C) a cylinder, (P) a prism.*

## *Mirror on the Wall*

The mirrors that hang on walls are usually flat ones and we are all very familiar with them. Figure 15-2 shows a scheme depicting how the rays of light reflected from a plane mirror give the impression that there is another similar object placed behind the mirror. But here is a question for those who think they know how a plane mirror works. The person you see in the mirror is very much like you, but there are a few oddities. He parts his hair on the wrong side and his tailor apparently made a mistake in placing the breast pocket on his coat; she wears her wedding ring on the wrong hand (or maybe she is from Europe), and you are rather sure that an X-ray picture of your mirror image would show your heart on the wrong side. In other words, a mirror turns things right to left and left to right. But on the other hand, the mirror on the wall does not turn up down and down up! The head is not turned into feet and the feet are not turned into a head. Why? To give another example, write T O M on a piece of paper, turn it toward a mirror, and you will read M O T. But if you write it in

T
Chinese fashion, O, and look at it in the mirror, you will
M

not notice any difference. Isn't this strange? Why does the mirror rotate the image around the vertical axis and not amount the horizontal one? Is it connected with the vertical direction of the force of terrestrial gravity? Or does it depend on the fact that our two hands are more similar to one another than our head and our feet? Or, maybe it is because we have two eyes located on a line perpendicular to the "head to feet" direction? (By closing one eye, you can easily disprove this latter possibility.)

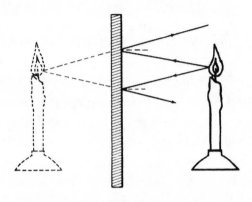

*Figure 15-2. The reflection of light from a plane mirror.*

The answer to this riddle is that the phenomenon does depend on the dissimilarity of the upper and lower parts of our body. Suppose you are walking along a street and hear an unexpected sound from behind. You turn around to see what caused it. But there are two ways of looking back, as illustrated in Figure 15-3. The "normal" way is to turn around the vertical axis, thus keeping your feet on the ground, which is geometrically just as good, is to stand on your head. The reason that you usually make the first choice is due, of course, to purely anatomical considerations. Returning to our mirror problem, we face a rather similar situation. The person in the mirror faces in the opposite direction, so that to make a comparison between ourselves and our mirror image we have to imagine ourselves turning by 180°. If we do it in the customary way, deepening our feet on the ground, we will find that our right and left hand will not fit those of our image. However, if we do it in the less conventional way and stand on our head, our head and feet will be mixed up instead. We leave it to the reader to apply the same principles to the T O M vs. M O T problem.

*Figure 15-3. Two ways of looking backwards.*

## Convex and Concave Mirrors

If the surface of the mirror is not a plane but has a regular shape such as a segment of a sphere, light rays from a point source will still intersect at a single point (focal point), but the position of this point will depend on the curvature of the surface. Figure 15-4*a* shows the path of rays in the case of a convex mirror, the reflected rays give the impression that the object is behind the mirror, but in the case shown the image is smaller than the actual object. In the case of a concave mirror (Figure 15-4*b*), reflected rays intersect in front of the mirror to produce the so-called "real image," which, in the case shown, is smaller than the original object and in addition is turned upside down. See if you can tell by means of the diagrams how the size and direction of an image in a concave and a convex mirror depend on the distance of the object from the mirror.

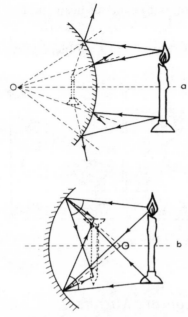

Figure 15-4. Formation of an image in convex (a) and concave (b) mirrors.

It should be noted here that in the case of spherical mirrors, the rays intersect in a single point only if the surface of the mirror corresponds to a small part of the complete sphere; otherwise, the focal point becomes somewhat blotted out and the image becomes blurred. This can be avoided, however, by using an elliptical mirror (Figure 15-5a), since, indeed, it can be proved geometrically that the lines connecting any point of an ellipsoid with two of its foci form exactly equal angles with the perpendicular at that point. Ellipsoidal mirrors are not often used in practice because for any given mirror of that kind, there is only one pair of points (the foci) where the object can be placed. If we keep one focus, $F_1$, in place and move another, $F_2$, into infinity, the ellipsoid becomes an open paraboloid (Figure 15-5b). The rays diverging from its focal point become parallel to each other after reflection, and, vice versa, parallel light beams falling on a parabolic mirror along its axis are collected exactly in its focal point.

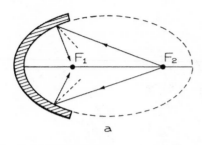

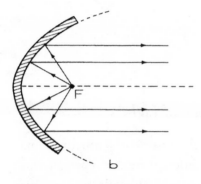

*Figure 15-5. The paths of rays in an elliptical (a) and a parabolic (b) mirror.*

This property of parabolic mirrors is widely used in the construction of various light-beam casting devices, such as headlights and searchlights, and in the making of astronomical instruments such as reflecting telescopes (or, simply, "reflectors") for obtaining sharp images of celestial bodies. There are three principle ways that astronomers can observe the images formed by the mirror of a reflecting telescope, as is shown in Figure 15-6 *a*, *b*, and *c*. In all three cases, a real image of the celestial object is formed by a large parabolic mirror and is inspected through an eyepiece similar in principle to an ordinary magnifying glass. For photographic work, the eyepiece and the human eye are replaced by a camera, and, in fact, big instruments such as that on Palomar Mountain are used more as "telecameras" than as "telescopes."

Figure 15-6. Three ways of viewing the image in a reflecting telescope.

## Refraction of Light

To study the refraction of light, instead of a mirror we may use semicylinders made of different transparent materials, which may be glass, plastic, or a thin plastic container filled with some liquid. We will find that the light that falls on a flat surface of the semi-cylinder and penetrates into it is *refracted*—i.e., is bent in a different direction from the original one. The angle formed by the refracted beam with the surface of the semi-cylinder depends on the corresponding angle for the incident beam and also on the material of the semi-cylinder. This is illustrated in the series of drawings given in Figure 15-7a, b, and c.

The law of refraction is somewhat more complicated than the law of reflection, but nevertheless it can be formulated in a comparatively simple way. If we mark on the directions of incident and refracted beams two points, A and C (Figure 15-8), equidistant from the entrance point,

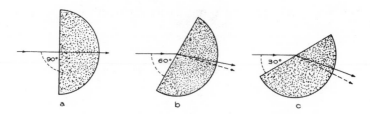

Figure 15-7. The refraction of light when the incident beams form angles of 90°, 60°, and 30°. Solid and broken lines correspond to red and blue light, respectively.

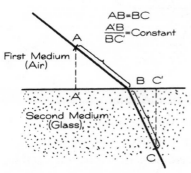

Figure 15-8. A demonstration of the Law of Refraction

B ($BA = BC$), and draw two perpendiculars from these points to the surface, thus fixing points $A'$ and $C'$, then **the ratio BA'/ BC' is independent of the direction of the incident light beam for any given pair of substances** (though it will have different values for different pairs of substances chosen for the experiment). This ratio is known as the *refractive index* of Substance II with respect to Substance I. In Table 15-1 we list the refractive indices of a few substances with respect to air, but since air differs very little in this respect from a vacuum, they can also be considered as "correct" refractive indices expressing the optical properties of these materials with respect to empty space. (In the same way the weight of an object measured in the air where Archimedes' Law tells us it will be buoyed up slightly by the weight of the displaced air is for all practical purposes the same as its "correct" weight in a vacuum.)

TABLE 15-1 REFRACTIVE INDICES OF DIFFERENT SUBSTANCES
(FOR YELLOW LIGHT)

| Substance | Refractive Index (yellow light) |
|---|---|
| Water | 1.33 |
| Alcohol | 1.36 |
| Glass (flint) | 1.65 |
| Diamond | 2.42 |

We see from this table, for example, that air-to-water and air-to-glass refractive indices are 1.33 and 1.65, respectively. What then will be the water-to-glass refractive index? It was found that **the refractive index on the surface between any two substances is equal to the ratio of their individual refractive indices.** Thus, the water-to-glass refractive index is 1.65/1.35 or 1.22.

If a transparent solid is submerged in a transparent liquid with the same refractive index, it becomes practically invisible, since the light rays are not refracted when they pass through the boundary between the two substances. This fact was used by H. G. Wells in his story, "The Invisible Man," which is about a man who managed to make his body transparent by reducing its refractive index to 1. But Wells overlooked, probably intentionally, one essential point: the invisible man would also be blind, since the lenses in his eyeballs would not form any image on the retina.

## *Why Is Light Refracted?*

Why do light rays change their direction when they travel from air into water or glass? This question has an important bearing on the problem of the nature of light. Sir Isaac Newton, who, besides his great achievements in mechanics, made many important contributions to optics, believed that a light beam represents a swarm of tiny material particles that are emitted from light sources and fly at high speed through space. He visualized the refraction of light rays entering any material medium as being caused by a certain attractive force (similar to ordinary surface tension forces) acting on the particles of light when they cross the surface of a material body (Figure 15-9a). The denser the material is, he thought, the stronger the force is and, hence, the greater the refraction. It is important to realize that, according to Newton's views, the velocity of light in substances with a higher refractive index should be larger than the velocity of light in air or in a vacuum, because the force pulling the light particles into the denser materials adds to their velocity.

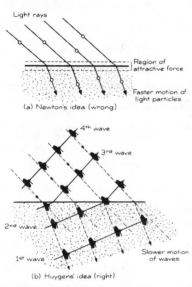

*Figure 15-9. Newton's and Huygens' explanations of the refraction of light.*

Newton's views were opposed by the Dutch physicist, Christian Huygens (1629-1695), who believed light to be a wave motion in a certain all-penetrating medium ("world ether") and the propagation of light to be similar to the propagation of sound waves through the air. It is instructive to notice that the explanation of the refraction of light based on Huygens' ideas leads to conclusions concerning light velocity in dense media that are directly opposed to those reached by Newton. To understand it in a simple way, let us substitute for successive wave of light approaching the surface of glass, successive waves of tanks operating in open country such as the North African desert. As the tanks cross a boundary between comparatively good terrain and poor, sandy terrain, their velocity is reduced (Figure 15-9*b*). If, after getting into the sandy terrain, individual tank commanders stubbornly maintain their original direction of motion, the advancing line of tanks that entered the sandy terrain earlier will be delayed in their advance in respect to those that enter it later. To make the rows of tanks continue to be at right angles to their direction of advance, it becomes necessary to change the course of advance and turn all the tanks somewhat to the right. It is questionable whether this argument would be acceptable to a Field Marshall Montgomery (or to a General Rommel for that matter), but this is exactly what light waves do. It is easy to conclude from the diagram in Figure 15-9 that, according to Huygens' view, the refractive index of the second medium with respect to the first is equal to the inverse ratio of the light velocities in them. Thus, the refractive index of glass with respect to water is:

$$\text{R.I. (water} \rightarrow \text{glass)} = \frac{\text{light velocity in water}}{\text{light velocity in glass}}$$

$$= \frac{\text{light velocity in water}}{\text{light velocity in vacuum}} \times \frac{\text{light velocity in vacuum}}{\text{light velocity in glass}}$$

$$= \frac{\text{R.I. (glass} \rightarrow \text{vaccuum)}}{\text{R.I. (water} \rightarrow \text{vaccuum)}}$$

which proves the empirical relation between refractive indices of different substances as stated earlier in the previous section.

## *Wave Nature of Light*

If at the time of the Newton vs. Huygens dispute physicists had known a way of measuring the velocity of light in different substances, the argument could have been settled in the simple way we have described. But, since this was not the case, the decision rested on another basic property of wave motion known as the interference of two light beams. We have already described the interference phenomenon in connection with the waves on the surface of a liquid (Figures 12-3 and 12-4), and the same reasoning can apparently be applied to the interaction of two light waves.

But how can we synchronize the vibrations of two light sources? Well, this can be done with mirrors, as was first suggested in 1814 by the French physicist, Augustin Fresnel (1788-1827). Fresnel's idea is illustrated in Figure 15-10. If we have a source of light $S$ and two mirrors, $M_1$ and $M_2$, forming a very small angle $\theta$ between them we will have two images, $S_1$ and $S_2$, located very close to each other. Since $S_1$ and $S_2$ are identical images of the original source

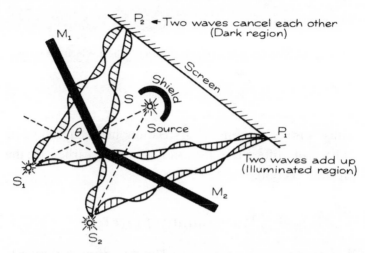

*Figure 15-10. Fresnel's experiment with the interference of two light beams.*

S, the light waves coming from them will be perfectly synchronized; in fact these two waves come actually from the single source S. Point $P_1$ is located on the screen in such a way that the difference between the distances $P_1S_1$ and $P_1S_2$ is equal to an integral number of wave lengths. In this case the two light waves will arrive at $P_1$ in phase, and the illumination at this point will be increased. On the other hand, at point $P_2$, located in such a way that $P_2S_2-P_2S_1$ is equal to an integral number of wave lengths plus one-half, light waves from $S_1$ and $S_2$ will arrive out of phase; the crest of one of them will always overlap the trough of the other, and as a result the illumination will be zero. We would therefore expect that the screen would be covered by a system of light and dark bands instead of being uniformly illuminated, and that is exactly what happens. This simple experiment proved beyond any doubt the correctness of Huygens' point of view and also provided a simple method of measuring the wave length of light. As we shall see later, the color of light depends on its wave length, and since light consists of a mixture of different wave lengths, the patterns seen on the screen will be somewhat blurred and will consist of rainbow colors. To eliminate this, we should place a piece of colored glass in front of the source of light, which will absorb all light except that of a single color (i.e., of a single narrow range of wave lengths).

The idea of the interference of water waves, as shown in Figure 12-5*b*, can be extended to a very long breakwater with many openings in it. This is analogous to the effect of an optical "grating" on light waves. We can make a "transmission grating" from a piece of glass on which we have scratched a large number of fine parallel grooves. When placed in a beam of light, the glass will transmit (i.e., let pass through) only the light that falls on the smooth unscratched strips between the grooves. A "reflection grating" works in a similar manner by reflecting light from the smooth sides of accurately shaped grooves in the surface of a metal mirror. With modern techniques, we can make gratings of several thousand scratches per millimeter. These optical gratings are a useful substitute for prisms in the study of optical spectra. Since the angular separation between separate beams depends on the wave length, white light falling on such gratings will be broken up into rainbow-colored strips.

The wave nature of light is also revealed when light passes through a very small opening that has dimensions comparable to the wave length of light (i.e., about 0.001 mm). If the opening is much larger than the wave length, light passing through it will make a spot of light the same shape as the opening on a screen placed behind it.

However, in the case of a small opening, the light will be scattered (diffracted), and all that will be seen on the screen will be a diffused, luminous spot from which we cannot conclude anything about the shape or size of the opening. This phenomenon, known as "diffraction of light," places a lower limit on the size of small objects that can be seen or photographed by using visible light. In fact, light waves cannot produce a picture of objects that are only the length of a light wave for the same reason that a painter cannot paint a miniature portrait using a two-inch brush.

Considering light as waves propagating through space, scientists logically assumed that there must be some medium through which these waves propagate. Since light propagates easily through empty space (which is not true in the case of sound), this hypothetical medium was assumed to fill all space and also to penetrate into the interior of all material bodies. It was called "light ether" or "world ether." We have already encountered this notion in a previous chapter in connection with Faraday-Maxwell's view on the nature of electric and magnetic fields, and we have also seen that electromagnetic waves were considered as the propagation of some kind of elastic deformation through this medium. In fact, light waves are electromagnetic waves, and differ from radio waves only by their very short wave length.

## *Light Emission by Hot Bodies*

We all know that in order to emit light, material bodies must be heated above a certain temperature. Hot radiators of a room heating system (below 100°C) do not emit any visible light whatsoever, while the heating units of an electric range (at about 750°C) glow with a faint reddish light that can be seen only if the kitchen is not too brightly illuminated. The filament of an electric bulb (about 2,300°C), which is much hotter than the kitchen range units, emits intense white light, while the light of an electric arc, which is still hotter (3,500°C), is almost uncomfortably intense and possesses a bluish tint. Thus, **the intensity of light that a heated body emits increases as its temperature increases, and the prevailing wave length shifts from the red toward the blue end of the spectrum.** The studies of these curves led to two fundamental laws of thermal radiation that may be formulated in the following way:

*Wien's Law*: The wave length of maximum intensity is inversely proportional to the temperature of the emitter.

*The Stefan-Boltzmann Law*: The total intensity of the emitted radiation is proportional to the fourth power of the emitter's temperature.

Thus, if we could increase by a factor of 2 the temperature of the filament in an electric bulb without melting it, the lamp would become considerably bluer and also 16 times brighter. These two laws of thermal radiation play a very important role in physics.

## Infrared and Ultraviolet Radiation

Just as in acoustical phenomena a *human ear can hear only the sounds* within a certain frequency (or wave length) interval, so the *human eye can see only the light within narrow limits of frequencies* (or wave lengths). Radiation with wave lengths longer than that of a red light is known as *infrared radiation*. It is also often called "heat radiation" since it is emitted from hot bodies (such as a room radiator) that are not yet hot enough to be luminous. In fact, heat rays are emitted by all material bodies no matter how low their temperatures are, but, according to the Stefan-Boltzmann Law, their intensity falls very rapidly with the temperature.

*Ultraviolet radiation* has wave lengths shorter than the blue-violet end of the spectrum, and this radiation becomes more and more important with the increasing temperature of the emitting body. While the ordinary electric bulb (at 2,300°C) does not emit any ultraviolet radiation to speak of, the Sun which has a surface temperature of about 6,000°C, emits enough ultraviolet radiation to sunburn or tan the exposed parts of the human skin. As an extreme case, there is a star located in the center of the so-called Crab Nebula that has a surface temperature of 500,000°C. At this tremendously high temperature, the prevailing wave length is shifted, according to Wien's Law, so far into the short-wave region that only a small fraction of its energy is emitted within the visible range. Most of the remaining energy is radiated in the invisible ultraviolet.

## *Line Spectra*

While the light emitted by hot solid bodies represents a mixture of all wave lengths and is known as a "continuous spectrum," gases behave, in this respect, quite differently. Take a candle and analyze the light from the flame by means of a simple spectroscope (Figure 15-11). You will see a continuous spectrum which looks about the same as the spectrum of light emitted by a hot solid body. But the fact is that in the case of a candle (or the flame in a fireplace), we do see the light emitted by solid bodies. It can be shown, indeed, that the main portion of light is emitted, not by the hot gas which forms the flame, but by tiny particles of unburned carbon (chimney black) that are heated to high temperatures by the flame gases. If the burning is complete, as it is in the case of a Bunsen gas burner, the flame loses most of its brightness.

If we introduce into the flame of a Bunsen burner a small amount of table salt, we will notice that the flame becomes a brilliant yellow, and if we look at it through a spectroscope, we will see, not a continuous rainbow-colored band, but a single narrow yellow line on a generally dark background. This line is produced by the sodium that is introduced into the flame. In fact, the line always appears when a substance introduced into the flame contains some sodium, and it is never present if the substance is lacking in that element. If, instead of a sodium compound, we put some lithium salt into the flame, the flame will become a brilliant red, while compounds of copper will give it a greenish tint. Another method of producing line spectra of different gaseous materials consists of subjecting them to an electric discharge, as is done, for example, in luminous advertising signs.

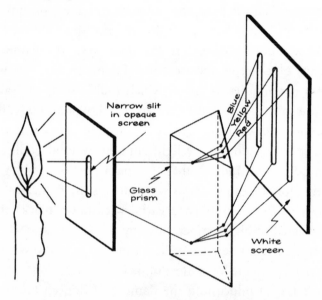

*Figure 15-11. The three lines shown on the screen at right are the particular colors (or wave lengths) of blue, yellow, and red. Each one of the colors present in the light of the flame will produce a similar line on the screen, and the infinite number of wave lengths from the incandescent carbon particles will make a continuous spectrum.*

The fact that each chemical element emits a set of spectral lines characteristic only of that particular element is the basis of so-called *spectral analysis*, i.e., the method by which we can observe the spectral lines emitted by a material of unknown chemical constitution and tell of what elements (and in what relative amounts) the substance is composed. Spectral analysis is particularly useful in astronomy for studying the composition of the Sun and the stars, since we clearly cannot get a piece of the material from these bodies to analyze by conventional methods in a chemical laboratory.

## Fraunhofer Lines

We can now continue our experiment with the table salt-colored flame by placing *behind* the flame a strong source of white light. While without the flame the source will produce a bright rainbow-colored continuous spectrum, the introduction of the sodium-contaminated flame between the white light source and the spectroscope will result in the appearance of a narrow black line exactly in the same place where the sodium emission line was previously found. The explanation of this fact is based on the notion of *resonance*. If the sodium atoms *emit* light waves of a certain characteristic frequency, they are also apt to *absorb* the waves of this particular frequency when an intense white light passes through them. The best-known example of such absorption lines is provided by the multitude of thin dark lines that cross the continuous spectrum of the Sun; these lines were discovered by the German physicist, Joseph Fraunhofer (1787-1826), and are therefore named after him. Fraunhofer Lines originate when the white light radiated by the deeper and denser layers of the solar body (the so-called "photosphere") passes through the thin gaseous envelope ("chromosphere") and "reversing layer" of the Sun. By observing them, we can make definite conclusions about the chemical composition of the Sun. The interesting fact revealed by this method is that, apart from the large predominance of hydrogen and helium gases, the chemical composition of the Sun is identical with that of our Earth. By using the methods of spectral analysis, astronomers have been able to learn a lot, not only about the Sun, but also about the chemical composition of the planetary atmospheres and the outer envelopes of distant stars.

## Why Solids Emit a Continuous Spectrum

Since each chemical element possesses a characteristic set of spectral lines, we must conclude that this characteristic frequency pattern is directly connected with the properties of atoms. We may visualize the different kinds of atoms as being similar to different kinds of musical instruments; an instrument emits a characteristic assortment of pure tones and the atoms of a chemical element emit a characteristic assortment of "pure colors," i.e., sets of spectral lines. When atoms are tightly piled together as in a solid or liquid body, they disturb each other so much that instead of a definite "optical chord," they produce nothing but an "optical noise" caused by the orderless superimposition of all possible frequencies. In fact, this situation is like having several dozen tuning forks shaking in a bag: lots of clinking and squeaking, but no pure tones. On the other hand, in gases the atoms fly freely through space and most of the time are too far from neighboring atoms to be disturbed by them. This gives them a chance to emit their characteristic frequencies between successive collisions.

## *Spectral Lines and Atomic Structure*

The study of the spectral lines that characterize various elements led to the discovery of a number of amusing regularities that are apparently due to the specific properties of atomic mechanisms responsible for the emission of these lines. In the case of hydrogen, which is the simplest chemical element, the relation between various lines shown in Figure 15-12 can be expressed by an extremely simple mathematical formula that was discovered in 1885 by the German school teacher, J. J. Balmer, and carries his name. This formula states that the frequencies of the observed lines are exactly proportional to the difference between the inverse square of 2 (i.e., 1/4) and the inverse squares of 3, 4, 5, etc. In fact, calculating these differences, we obtain:

$$\left(\frac{1}{2^2} - \frac{1}{3^2}\right) = \left(\frac{1}{4} - \frac{1}{9}\right) = 0.138889$$

$$\left(\frac{1}{2^2} - \frac{1}{4^2}\right) = \left(\frac{1}{4} - \frac{1}{16}\right) = 0.1875$$

$$\left(\frac{1}{2^2} - \frac{1}{5^2}\right) = \left(\frac{1}{2} - \frac{1}{25}\right) = 0.21$$

$$\left(\frac{1}{2^2} - \frac{1}{6^2}\right) = \left(\frac{1}{4} - \frac{1}{36}\right) = 0.222222$$

etc.

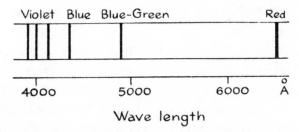

*Figure 15-12. Part of the hydrogen spectrum with the color of different emission lines indicated.*

Multiplying these values by $3.28937 \times 10^{15}$ we obtain:

$$4.5686 \times 10^{14}$$
$$6.1676 \times 10^{14}$$
$$6.9077 \times 10^{14}$$
$$7.3097 \times 10^{14}$$
etc.

which is to be compared with the values:

$$4.5685 \times 10^{14}$$
$$6.1676 \times 10^{14}$$
$$6.9075 \times 10^{14}$$
$$7.3099 \times 10^{14}$$
etc.

representing the actually measured frequency of hydrogen's spectral lines. The agreement is good up to the fifth decimal place.

The line spectra of alkali metals, such as lithium, sodium, and potassium, are similar to the spectrum of hydrogen, through somewhat distorted, in dictating that the structure of these atoms is similar to that of hydrogen. But in other elements, as for example in oxygen or iron, the pattern of spectral lines becomes so very complicated that any mathematical formulation becomes extremely difficult.

For some time after the discovery of characteristic spectral lines and the regularities existing between them, physicists tried to explain these regularities by considering atoms as miniature elastic bodies of various shapes. Thus, a hydrogen atom was considered to be a sphere, a sodium atom an elongated ellipsoid, and an oxygen atom a doughnut with a very small central hole. The physicists hoped that they could calculate the vibration frequencies of these variously shaped bodies and find that they were in agreement with the observed sets of spectral lines. However, all these attempts resulted in a fiasco; the picture of atoms as miniature vibrating bodies was too simple and too naive. A complete understanding of the laws of optical spectra came only much later as the result of exhaustive studies of the inner electric structure of matter.

## The Doppler Effect

An important phenomenon, both in the case of light waves and of sound waves, is the change of wave length in the light or sound that comes to us from a moving source. This phenomenon is known as the Doppler Effect. If the source of a given frequency moves toward us, the waves coming in our direction will be squeezed and will appear to us shorter than those from a stationary source. In the case of a receding source, the situation is the opposite, and the arriving waves will be longer. How it happens can be easily understood by inspecting the diagram in Figure 15-13.

In the case of sound, the approaching source will appear to have a higher pitch and the receding source a lower pitch. In the case of a moving light source, we will see violet and red shifts of spectral lines. We can show that the relative change of wave length $\Delta\lambda/\lambda$ due to the Doppler Effect equals the ratio of the velocity of the source to the propagation velocity of waves (sound or light velocity). Thus, by measuring the change of wave length, we can easily calculate the velocity of the source. This method is widely used in astronomy for estimating the radial velocities of stars and galaxies from their observed spectra.

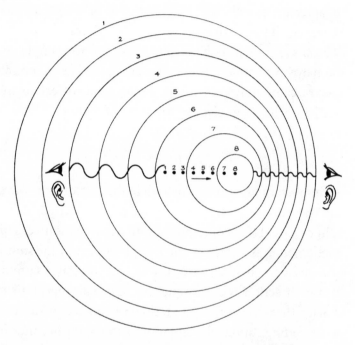

*Figure 15-13. The consecutive positions of spherical waves emitted by a source moving to the right. The positions of the source and the corresponding waves are indicated by the same numbers.*

The Doppler Effect also exists, of course, in the case of a stationary source and a moving observer, and there is a story (not a true one, of course) that the famous American physicist, R. W. Wood, once tried to use this effect in a traffic court as an excuse for going through a red light. The judge, so the story goes, was almost ready to let the famous man go without paying the fine when one of Wood's students, whom he had recently flunked on an examination in optics, proposed that the judge ask the professor to calculate the velocity with which he must have been driving toward the red signal light to see it green. As a result, Wood had to pay a much higher fine for greatly exceeding the speed limit of the city of Baltimore.

## *Polarization of Light*

We saw in Chapter 12 that there are two possible ways of wave propagation: (1) longitudinal waves in which the motion of individual particles is along the line of propagation, and (2) transverse waves in which the motion is perpendicular to the line of propagation. Which kind of wave motion do we have in the case of light? The important difference between the longitudinal and transverse waves is that the latter can be "polarized." To understand this important notion, let us look at a wave in the direction of its propagation as shown in Figure 15-14. In the case of longitudinal waves, (a), the motion of the particle takes place perpendicular to the surface of the paper and will not be noticeable from the direction we are observing it. In the case of transverse waves (b and c), the motion of the particles is in the plane of the paper and easily observable in that projection. We call the transverse wave "natural" or "nonpolarized" if the motion of the particles takes place in all possible directions (b); if the motion is in only one direction (c), the wave is "polarized." The notion of polarization can be clarified by the analogy given in Figure 15-15. Suppose we have a sieve, made of a set of parallel wires without cross wires, and drop matches on it in such

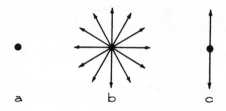

*Figure 15-14. Cross-section of a wave in the direction of its propagation.*

a way that the falling matches remain horizontal but may have different orientations in the horizontal plane. It is clear that only those matches that are parallel to the sieve wires will pass through. If under this "match polarizer" we place a similar "match analyzer," the "beam of matches" will pass through only if the wires in the lower sieve run parallel to those of the upper sieve. We can extend this analogy to the case of longitudinal waves by dropping the matches in a vertical position. In this case, all the matches will pass through, regardless of the relative positions of the two sieves.

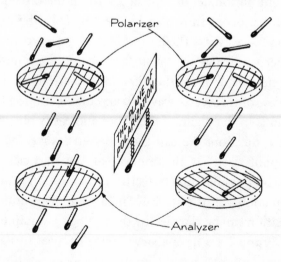

*Figure 15-15. "Polarized matches."*

It was found that many crystals have the ability to polarize light waves in such a way that the light is able to pass through a pair of crystals when their "optical axes" (sieve wires in Figure 15-15) are parallel to one another but is completely stopped when the axes are crossed. This fact shows beyond any doubt that *the vibrations of light waves are perpendicular to the propagation of motion.*

We cannot finish this already overgrown chapter on light without mentioning the phenomenon known as "rotation of the plane of polarization." If we place certain crystal objects in a beam of polarized light between a polarizer and an analyzer, we will notice that the plane of polarization has been rotated, and in order to extinguish the passing light we have to turn the analyzer by a certain angle. Properly speaking, the passing light can be completely extinguished only if we use a monochromatic source, since the rotation of the plane of polarization depends on the wave length of light in very much the same way as it does in the phenomenon of refraction. If we place crystals or other materials, such as certain plastics (crumpled cellophane will do nicely) in the beam of white polarized light, we will observe beautiful rainbow patterns that are very helpful in the study of crystalline structures.

# CHAPTER 16

# Modern Views on Space, Time, and Motion

In discussing the phenomena of electromagnetism and light, we referred time and again to the notion of "world ether," the hypothetical all-penetrating substance that was supposed to be responsible for long-range electric and magnetic interactions between material bodies as well as for the propagation of light waves across what we usually consider to be empty space. In order to carry out all these functions, "world ether" had to be a solid medium, since only a solid body can be subjected to elastic stresses and since only in a solid medium can transverse elastic waves exist. The idea of a solid material filling the entire space of the universe without, however, offering any resistance to the motion of material bodies passing through it naturally led to grave conceptual difficulties. These difficulties finally culminated in the paradoxical result of an experiment carried out especially for the purpose of detecting the motion of our Earth through this hypothetical universal substratum. Since this experiment pertains to the effect of the Earth's motion through space on the observed velocity of light, we start with a description of the methods used for measuring that velocity.

## Velocity of Light

The first attempt to measure the propagation of light was undertaken by Galileo in a very primitive way. One evening, he and his assistant placed themselves on two distant hills in the neighborhood of Florence, each of them carrying a lantern with a shutter. Galileo's assistant was instructed to open his lantern as soon as he noticed the flash from the one carried by his master. If light was propagating with a finite speed, the flash from the assistant's lantern would have been observed by Galileo with a certain delay. The result of this experiment was, however, completely negative, and we now know very well why. Light propagates so fast that the expected delay in Galileo's experiment must have been about one hundred-thousandth of a second, which is quite unnoticeable to human senses.

The first successful measurement of the velocity of light was carried out in 1965 by the German astronomer, Roemer, who replaced Galileo's assistant by the moons of the planet Jupiter, thus increasing the distance to be covered by light by a factor of hundreds of millions.* Roemer's method is illustrated in Figure 16-1, which shows the orbits of the Earth, Jupiter, and one of its moons. Moving around the planet, the moons are periodically eclipsed as they enter the broad cone of shadow cast by Jupiter. Studying these eclipses, Roemer noticed that sometimes they took place as much as eight minutes ahead of schedule and sometimes with a delay of eight minutes. He also noticed

---

*It may be remarked here that Galileo still had a hand in Roemer's measurement of the velocity of light, since it was he who discovered the moons of Jupiter.

that the eclipses were early when the Earth and Jupiter were on the same side of the Sun (1st position) and delayed in the opposite case (2nd position). Ascribing correctly the observed irregularities to the difference of time taken by light to cover the changing distance between the Earth and Jupiter, Roemer calculated that light must be propagated through space at a speed of about 300,000 km/sec ($3 \times 10^{10}$ cm/sec).

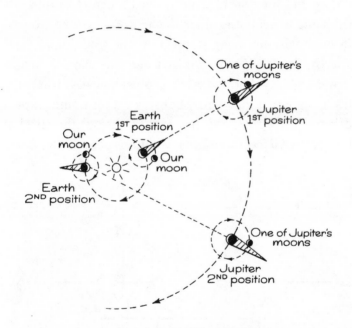

Figure 16-1. Roemer's method of measuring the velocity of light by observing the eclipses of Jupiter's moons.

The first laboratory measurement of the speed of light was carried out in 1849 by the French physicist, H. L. Fizeau (1819-1896), whose apparatus is shown in Figure 16-2.

It consists essentially of a pair of cogwheels set at the opposite ends of a long axis. The wheels were positioned in such a way that the cogs of one were opposite the intercog openings of the other, so that a light beam from the source could not be seen by the eye no matter what the position of the wheels. However, if the wheels were set in fast rotation and the speed of this rotation was such that the wheels moved by half the distance between the neighboring cogs during the interval of time taken by light to propagate from one wheel to the other, the light was expected to pass through without being stopped. As the reader must have already noticed, Fizeau utilized here the same principle that is often used nowadays on express highways where the traffic signals at the intersections are set for uninterrupted driving at a legal speed. In order to observe the effect at the speed of a few thousand revolutions per minute, which

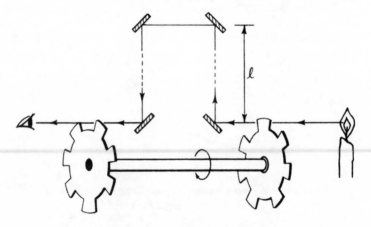

Figure 16-2. The Fizeau method for measuring the velocity of light.

was about the maximum that Fizeau could achieve, he had to lengthen the path of the light beam by using four mirrors as shown in Figure 16-2. This direct laboratory measurement gave a value for the speed of light that stood in reasonably good agreement with that obtained by Roemer's astronomical method.

## *Ether Wind?*

If it were true that light waves propagate through a jelly-like "world ether" which fills universal space, we would be able to notice our motion through space by observing the effect of that motion on the velocity of light. In fact, since the Earth moves along its orbit at a speed of 30 km/sec, we would experience an "ether wind" blowing in the direction opposite to our motion in the very same way that a speeding motorcyclist experiences a strong "air wind" blowing into his face. Light waves propagating in the direction of that "ether wind" would move faster, being helped by the motion of the medium, while those propagating in the opposite direction would be slowed down. In the year 1887, the American physicist, A. A. Michelson (1852-1931), carried out an experiment that was expected to demonstrate the effect of the Earth's motion on the velocity of light as measured on its surface. Instead of measuring the velocity of light in two opposite directions, Michelson found it more convenient to measure it in two mutually perpendicular directions.

In order to understand Michelson's scheme, let us consider a Mississippi steamboat running between St. Louis, Missouri, and Memphis, Tennessee, some 300 miles apart. Sailing downstream the boat moves faster, since it is helped by the current, while on the way back it is correspondingly retarded. Does the gain in time one way compensate for the loss in time the other? Although it may appear at first sight that it does, this conclusion is not true. Let us do some simple arithmetic, assuming that the boat's speed (in still water) is 30 mph and that the velocity of the stream is 3 mph. Sailing up and downstream, the boat will have the velocities, relative to the shore, of 27 and 33 mph. The time necessary for the round trip apparently will be:

$$\frac{300}{27} + \frac{300}{33} = 11.11 + 9.09 = 20.20 \text{ hours}$$

which is 1 percent more than it would be in still water. The closer the velocity of the stream is to the velocity of the boat, the longer is the time necessary for the round trip, and if the two velocities are equal the boat will never return!

Let us consider now the problem of how to move across the stream as it confronts a ferry connecting two ends of a highway at two opposite points across the river (Figure 16-3). It is clear that when crossing the river the ferry must keep its course slightly upstream to compensate for the drift. Thus, while it covers the distance $AB$ in respect to the water, it drifts downstream by the distance $BC$. Clearly the ratio $BC/AB$ is equal to the ratio of stream and ferry velocities, and we will take it, as in the previous example, to be equal to 1/10. Applying the Pythagorean Theorem to the rectangular triangle $ABC$, we can write:

$$(AC)^2 + \left( AB\frac{\text{river velocity}}{\text{boat velocity}} \right)^2 = (AB)^2$$

or:

$$(AB)^2 \left[ 1 - \left( \frac{river\ velocity}{boat\ velocity} \right)^2 \right] = (AC)^2$$

Since we have assumed that the ration of velocities is 1/10, we find from the above formula that:

$$AB = \frac{AC}{\sqrt{1 - 1/100}} = 1.005 \times AC$$

Since this result applies equally well to both crossings, the distance, with respect to the water to be covered by our motor launch on a round trip across the river, will also be 0.5 percent longer and so will be the time needed for the trip. Thus we find that moving across the stream and back also introduces a delay, but this delay is only half as much as the delay connected with sailing up and downstream.

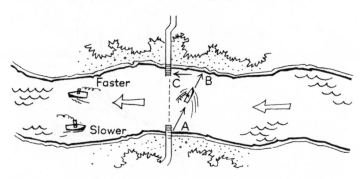

*Figure 16-3. How to propel a ferry across a moving stream.*

Now substitute the "ether wind" for the river, and propagating light waves for the boats, and you will have the principle of Michelson's experiment. The details are shown in Figure 16-4. A light beam from a source, *S*, falls on the glass plate, $P$, covered with a thin semi-transparent layer of silver which reflects half of the beam in the direction of the mirror, $M_1$, and passes the other half through in the direction of mirror $M_2$. Being reflected by these mirrors, the beams return to the plate, $P$; half of the first beam penetrates the thin silver coating on $P$ and continues on to the telescope and the eye at $T$; half of the second beam is reflected into the telescope by the silver layer. Thus, the two beams entering the telescope will have the same intensity and the same phase, so that, according to the considerations in Chapter 15 the field of vision will be brightly illuminated.

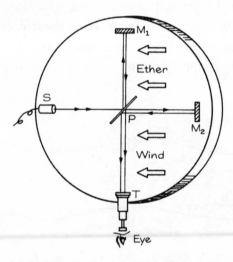

*Figure 16-4. Michelson's apparatus. All instruments were installed on a round marble plate floating in mercury, so that the system could be rotated without shaking. The separate light beams are shown here somewhat apart from one another in order to make the drawing clearer.*

However, in the presence of "ether wind," the situation was expected to be rather different. When the apparatus was placed in such a way that the line $PM_2$ coincided with the direction of the wind or its projection on a horizontal plane, the light waves traveling in this direction would be in the position of a boat sailing up and downstream, whereas the light waves traveling along the line $PM_1$ would correspond to a ferry moving to and fro across the river. Because of the difference in time delays in the two cases, the light beams would not arrive at the telescope simultaneously, and the ensuing difference in their arrival would result in interference which would reduce the brightness of the field of vision. The ratio of the orbital velocity of the Earth, 30 km/sec, to the velocity of light, 300,000 km/sec, is considerably smaller than in the previously discussed nautical example. Using the same method of calculation, we find that in this case the two light beams should arrive at the telescope with a relative delay of only $5 \times 10^{-7}$ percent of the total travel time. If (which is about right) the distance between the central plate of Michelson's apparatus and the mirrors is 150 cm, the total travel time (plate to mirror and back) is

$$\frac{300}{3 \times 10^{10}} = 10^{-8} \text{ sec,}$$

so that one light beam should be delayed in respect to the other by $5 \times 10^{-17}$ sec. Although this is a very short time from our everyday point of view, it is fairly long from the point of view of wave optics. Indeed, during that time interval, light propagates by a distance of $5 \times 10^{-17} \times 3 \times 10^{10} = 1.5 \times 10^{-6}$ cm, which is about 5 percent of the wave length.

Turning the apparatus by 90° and thus exchanging the roles of the mirrors, $M_1$ and $M_2$, we would expect the same delay in the opposite direction. Thus the total difference between the two light beams in the first and in the second positions of the apparatus was expected to be 10 percent of the wave length and should have caused an easily noticeable change in the illumination of the field of vision in the telescope.

However, to his great surprise, and to the surprise of the entire scientific world (at least in the fields of physics and astronomy), Michelson failed to notice any change at all! How could it be? Some scientists suggested that there might be some drag of the "world ether" caused by the moving bulk of the Earth so that the resulting velocity of the "ether wind" near the ground is considerably reduced. However, the repetition of Michelson's experiments carried out on top of a mountain and in a high-flying balloon disproved that possibility. A British physicist, G. F. Fitzgerald (1851-1901), tried to interpret the negative result of Michelson's experiment by postulating that all material bodies moving through the "world ether" shrink in the direction of their motion by an amount dependent on their velocity. The effect of this so-called Fitzgerald Contraction would be to transform the round table on which Michelson's mirrors were mounted into an ellipse with the shorter axis in the direction of the Earth's motion. This would reduce the distance to be traveled by the light beam propagating in the "up and down wind" direction and would enable that light beam to arrive at the telescope simultaneously with the beam that was traveling across the wind. Numerous attempts were made to explain this hypothetical contraction by the change of electric and magnetic forces between the atoms moving through the "world ether," but they never led to any positive result.

## Conceptual Failure of World Ether Notion

Michelson's failure to detect the motion of the Earth through the "world ether" had the same roots as the failure of contemporary physical theories to formulate the mechanical properties of this hypothetical medium. As we have already discussed earlier in some detail, it was illogical to ascribe to the hypothetical "world ether" the properties of ordinary matter, such as, for example, elasticity or compressibility, since in doing so we would also have to assume that "world ether" possesses some kind of granular structure formed by "sub-atoms." But if, on the other hand, we consider "world ether" to be an absolutely homogeneous substance without any internal structure, there is no logical possibility of talking about the motion of that ether or the motion of objects in respect to it. In fact, when we watch a rotating disc we notice that it rotates by observing the motion of minor marks on its surface, such as scratches or dents. If the surface of the disc is perfectly smooth, with no marks that would catch our eye, we will not be able to tell just by looking at it whether it is moving or not. But, of course, we can touch it with our finger tip and immediately feel whether its surface is at rest or slides under our finger. And if the disc rotates fast enough, we will feel the warmth produced by the friction between the skin of our finger and the moving surface of the disc. But the phenomenon of friction, which informs us about the state of motion of the disc, is again a purely molecular phenomenon and would be absent in an "absolutely homogeneous" substance.

If we give a little thought to the problem, we can easily persuade ourselves that **it is meaningless to talk about the motion of a continuous medium or the motion in respect to it unless this medium can be considered to be formed by individual discrete particles.** If we had a long ribbon made of an absolutely continuous material and observed a wave propagating along it (Figure 16-5), it would be meaningless to ask whether (a) it is a regular elastic wave propagating along a stationary ribbon, or (b) a rigid ribbon cut out of a sheet shaped like corrugated iron that is moving bodily from left to right.

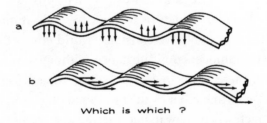

*Figure 16-5. (a) A wave propagating through a stationary elastic ribbon; (b) a moving wave-shaped rigid ribbon.*

## So Spoke Einstein

In the year 1905, Albert Einstein (1879-1955), who was at that time working as a patent clerk in Zurich and had just invented a new type of oil pump, looked at Michelson's failure to notice any motion through the "world ether" in a much more radical way than did his contemporaries. Instead of trying to patch up the accumulating difficulties and contradictions connected with the notion of "world ether," he rejected the notion outright as unsuitable for the description of the physical world and returned to the pre -etherian idea of a completely empty space. Along with the exit of "world ether" from the stage of physics, out went also the notion of "absolute motion" through space which was always associated, though often subconsciously, with the idea of motion with respect to "world ether." If there is no ether filling entire space and serving as a universal reference system for the motion of material bodies, **we can speak only about the motion of one material body relative to another material body, and the basic laws of physics should be the same no matter in which system of reference we are studying them.** Because of this basic postulate stating that there is no such thing as *absolute motion* and that only a *relative motion* of one object in respect to another has a physical meaning, Einstein's theory is commonly known as the *Theory of Relativity.*

It follows from the above postulate that it should be impossible to detect the motion of one system of reference in respect to another by performing some physical experiment in each of them and then comparing the results. Thus, the grandfather's clock in the captain's cabin on the *Queen Mary* speeding toward New York across the smooth, blue waters of the Atlantic with no storm or choppy sea breaking the uniformity of motion will operate just as well as if it were standing in the sitting room of his home. And the passengers playing ping-pong or billiards on this ship will not be able to tell whether their ship is lying quietly in a Southampton dock, or sailing across the blue Atlantic. Michelson's experiment had shown that this is also true for light phenomena. A physicist repeating Michelson's experiment in an outside cabin of the ship will not know whether the ship is moving or resting (relative to Great Britain) unless he goes up on deck and sees the gray buildings of dock installations or the limitless expanses of the ocean.

All this is very well. But what about the nature of electromagnetic interactions between material bodies, and what about the light waves propagating through empty, interstellar space? If there is no "world ether," then *what* attracts or repels the poles of two magnets, and *what* is oscillating when light or radio waves propagate through empty space? As has already been hinted in the two previous chapters, the rejection of "world ether" necessitates the introduction of a new physical entity, which is the electromagnetic field itself. *Instead of considering electric and magnetic fields as the stresses in a certain universal substratum, we now ascribe to them a definite physical reality—in fact, about as much reality as we ascribe to ordinary material bodies.*

Probably the most important innovation brought by Einstein to our views concerning electromagnetic and optical phenomena is that we must ascribe a certain mass to electromagnetic energy as well as to any other form of energy. Thus, a magnetized iron bar is slightly heavier than it was before magnetization, the difference being due to the weight of the magnetic field surrounding it. Similarly, a flashlight sending out a beam of light gradually loses weight because a certain amount of mass is being carried away by light waves. According to *Einstein's Law of the Equivalence of Mass and Energy*, **the mass (in grams) to be ascribed to a certain amount of energy (in ergs) is equal to that amount of energy divided by the square of the velocity of light (in cm/sec)**. Since the square of the velocity of light ($9 \times 10^{20}$) is a very large number, the mass of various forms of energy is usually very small. Thus, in the case of an ordinary laboratory magnet, the energy stored in the magnetic field surrounding it is about $10^6$ ergs, which corresponds to a mass of only

$$\frac{10^6}{9 \times 10^{20}} = 10^{-15} \, \text{gm.}$$

A flashlight with a 10-watt bulb emits $6 \times 10^9$ ergs/min, so that in each minute of operation the flashlight becomes lighter by $7 \times 10^{-12}$ gm. In astronomical applications, Einstein's law leads to considerably larger masses. For example, the Sun emits $4 \times 10^{11}$ tons of heat and light per day.

Thus, whereas according to the old-fashioned ideas, electric and magnetic fields were considered as elastic deformations in the all-penetrating "world ether" (Figure 16-6*a*), we now consider them as independent physical entities that possess a certain mass, have no internal granular structure, and surround electrically charged and magnetized objects, thinning out to zero away from them (Figure 16-6*b*). Similarly, while classical physics considered light waves as the propagation of elastic deformation through the "world ether" filling entire space (Figure 16-6*c*), we consider them now as vibrations taking place within the lumps of a certain physical entity (i.e., *electromagnetic field*), flying freely through the empty space (Figure 16-6*d*).

In other words, the propagation of an electromagnetic wave is more similar to the wave-like motion of a snake crawling through the grass and carrying along its body as well as its form of motion than it is to the waves on the surface of the water where only the form of motion but not the material itself is moving forward.

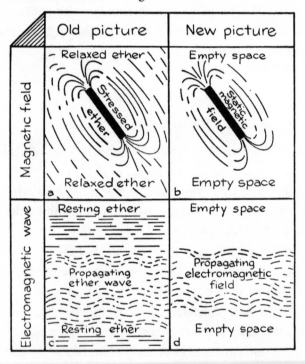

*Figure 16-6. The contrast between the pre-Einsteinian and the post-Einsteinian views of a magnetic field and an electromagnetic wave.*

Einstein's views concerning the close similarity between ordinary material bodies and electromagnetic fields have found a brilliant confirmation in more recent experimental findings, which have shown that under certain conditions electromagnetic fields can be transformed into material particles, and vice versa. In fact, gamma rays, which are short electromagnetic waves emitted by radioactive materials, can be turned into swarms of particles (positive and negative electrons), while, on the other hand, some material particles, such as the so-called neutral pions, can completely vanish by being transformed 100 percent into oscillating electromagnetic fields.

With these new views concerning the propagation of electromagnetic waves, and in particular that of light, we can return to the interpretation of the negative result of Michelson's experiment. There is no conversation now about "ether wind," and the light beams in Michelson's apparatus should be considered as vibrating lumps of electromagnetic fields flying more or less like rubber balls through space and being reflected by the mirrors and by the half-silvered plate. As would be the case with ordinary rubber balls, their arrival at the joining point $T$ of Michelson's apparatus will not be influenced at all by the motion of the apparatus, provided this motion is smooth and nonaccelerated. No matter whether Michelson's experiment is performed on the *Queen Mary* lying at anchor at the Southampton docks or sailing toward New York across the Atlantic, or on the Earth speeding on its orbit around the Sun, or on an interplanetary ship "anchored" in space with respect to fixed stars, it will always show the same negative result.

In a way, this new view concerning the nature of light represents a compromise between Newton's idea of a stream of particles moving along the light rays and Huygens' idea of waves propagating through the "world ether." In fact, it combines the best features of both theories.

## Relativistic Mechanics

As we saw in the previous section, the difficulties connected with Michelson's experiment have been resolved by abolishing the picture of light waves as propagating through the stationary "world ether" and replacing it with a picture of a vibrating material substance (electromagnetic field) moving bodily through empty space. However, this change in our way of considering the nature of light was still far from sufficient to straighten out all the difficulties pestering physics at the turn of this century. One trouble arose in connection with the velocity of light emitted by moving sources. If a rifleman sits in a fast-moving open-top automobile and shoots in a forward direction, the velocity of the bullets with respect to the ground will be the sum of the muzzle velocity and the velocity of the auto (Figure 16-7a), but bullets shot backwards will fly correspondingly slower (Figure 16-7b). If we consider light as some kind of vibrating bullets emitted by light sources, we would expect that the velocity of light emitted by an approaching source would be higher than that emitted by a receding source. Abundant astronomical evidence based on the observation of binary stars proves, however, beyond any doubt that this is not the case. A binary star (Figure 16-7c) is a system of two giant suns rotating around their common center of gravity and is a rather common object in the sky (in fact, about half of all known stars are binaries). Because of its rotation around the common center, each of the stars is moving toward us during one-half of its rotation period and away from us during the other half.

*Figure 16-7. The velocity of a bullet is affected by the motion of its source. Is this also true in the case of light?*

If the velocity of light were affected by the motion of the source, the light from the approaching star would reach us sooner than the light from the receding one, and the difference in the arrival of the two light signals would be quite large. Assuming, for example, that the orbital velocities of the two stars are the same as the orbital velocity of the Earth, i.e., 30 km/sec (and they are often larger than that), we find that the light would be accelerated or retarded by 0.01 percent depending on whether it comes from the approaching or from the receding component of a binary star. Over a distance of 100 light-years, which is not uncommon for the observed binaries, this seemingly small difference in velocity would result in one week's difference between the arrival of light from these two stars to the Earth, and this difference would be reversed every half revolution period. Thus, an astronomer observing a binary star would find himself in the position of a sports fan watching a prize fight on a TV screen which, because of some trouble in transmission during the third round, shows the champion and the challenger with a few minutes difference in phase. Our fan would see the champion *already* resting in his corner while the challenger is *still* shadow boxing in the middle of the arena, and a minute later the champion would go out for the kill while the challenger is *still* being readied by his aids. In the middle of the fourth round, the fight would seem to be normal but toward the end of it things would change in the opposite direction, and the surprised sports fan would see the champion deliver his K.O. blow after the challenger was already counted out. Since nothing of this kind has ever been seen by astronomers observing the motion of binary stars, we must conclude that **the velocity of light is not affected by the motion of its source.**

But this means that **if we add to the velocity of light any other velocity we get again the same original velocity of light!** This is paradoxical! It contradicts common sense! Well, said Einstein, if there is a scientifically established paradox you cannot get rid of, all you can do is rationalize it. And, as to the common sense...well, the same common sense was once objecting to the idea that the Earth is round. If the common-sense idea concerning the addition of two velocities does not apply to the velocity of light and the velocity of its source, it must be generally wrong and its common use in everyday life may be justified only by the fact that all the velocities we encounter in ordinary life are much smaller than that of light. Thus, cutting another Gordian knot, Einstein introduced a new and at first sight very fancy law governing the addition of two velocities. If $v$ is the velocity of the aforementioned automobile and $V$ the muzzle velocity of the bullet shot in the forward direction by the rifleman in the auto, the velocity of the bullet with respect to the ground will be, not $V + v$, but:

$$\frac{V + v}{1 + \dfrac{V \times v}{c^2}}$$

where $c$ is the velocity of light. If both velocities, $V$ and $v$, are small compared to the velocity of light, the second term in the denominator is practically zero, and the old "common sense" formula holds. But if either $V$ or $v$, or both, approaches the velocity of light, $c$, the situation will be quite different.

Suppose that the velocity of the auto is 75 percent of the speed of light and that the muzzle velocity of the rifleman's bullet is the same. According to common sense, the velocity of the bullet with respect to the ground should be 50 percent above the velocity of light. However, putting $V = 0.75\,c$ and $v = 0.75\,c$ into the above formula, we get only $0.96\,c$, so that the velocity of the bullet with respect to the ground remains less than the speed of light. The reader can easily verify this fact, that no matter how close the two velocities to be added are to the velocity of light, the resulting velocity will never exceed it. In the limiting case, if we make $v = c$, we obtain:

$$\frac{V+c}{1+\dfrac{V \times c}{c^2}} = \frac{V+c}{1+\dfrac{V}{c}} = \frac{c\,(V+c)}{(c+V)} = c$$

This is why the velocity of the source does not add anything to the velocity of light emitted by it. Fantastic as it may look at first sight, Einstein's Law for the Addition of Two Velocities is correct and has been confirmed by direct experiments. It does not agree with common-sense conclusions, but we should not forget that common-sense conclusions are based on our everyday experience, and neither an auto traveling with a speed close to that of light nor rifles shooting bullets at that speed can be considered as an "everyday experience"! Thus, Einstein's Theory of Relativity leads us to the conclusion that **it is impossible to exceed the velocity of light by adding two (or more) velocities, no matter how close each of these velocities is to that of light**. The velocity of light, therefore, assumes the role of some kind of *universal speed limit* which cannot be exceeded no matter what we do.

Another way to understand the existence of an upper limit for velocity is to consider the amount of energy that would be necessary to accelerate a material body to the velocity of light. As was discussed in Chapter 11, the kinetic energy of motion in classical mechanics is defined as half of the mass times the square of the velocity. Thus the energy of an object moving with the speed of light would be: $^1/_2$ mass $\times c^2$. However, according to Einstein's

Law of the Equivalence of Mass and Energy, we have to revise this conclusion by taking into account the fact that the kinetic energy of motion, as for any other form of energy, possesses a certain mass. Therefore the argument will run as follows: If a mass $M$ moves with the velocity of $v$, the kinetic energy of motion is $1/2\ Mv^2$. But since this kinetic energy possesses the mass $(1/2\ Mv^2)/c^2$ and since it moves along with the object, there must be an additional kinetic energy:

$$\frac{1}{2}\left(\frac{\frac{1}{2}Mv^2}{c^2}\right)v^2$$

However, this additional kinetic energy has the mass:

$$\frac{1}{2}\left(\frac{\frac{1}{2}Mv^2}{c^2}\right)\frac{v^2}{c^2}$$

and, moving with the velocity, $v$, must give a further contribution,

$$\frac{1}{2}\left[\frac{1}{2}\left(\frac{\frac{1}{2}Mv^2}{c^2}\right)\frac{v^2}{c^2}\right]v^2$$

to the total kinetic energy. But this additional kinetic energy possesses the mass, etc., etc. It can be shown that this roundabout approach finally adds up to give a fair approximation to:

$$M \times c^2\left(\frac{1}{\sqrt{1-\frac{v^2}{c^2}}}-1\right)$$

which represents the correct relativistic expression for the kinetic energy of an object moving with the velocity $v$. If the velocity, $v$, approaches the velocity of light, $c$, the expression under the radical in the above formula tends to zero, so that the kinetic energy of the moving object tends to infinity. Therefore we conclude that **it is impossible to accelerate a material object to the velocity of light (not to mention super-light velocities) because, in order to do this, we would need an infinite amount of energy.**

This aforementioned formula for kinetic energy has been confirmed by direct experiments on fast-moving electrons, and gives us another aspect of the relativistic postulate that considers the velocity of light as the maximum possible velocity.

## Space-Time Transformation

Einstein's new law for the addition of velocities clearly contradicts the classical (common sense) ideas concerning space and time, so that in accepting this new law as experimental fact we are forced to introduce radical changes in our old notions. In his previously mentioned monumental work, *Principia*, the great Newton wrote:

I. Absolute, true, and mathematical time, of itself, and from its own nature, flows equably without relation to anything external.

II. Absolute space, in its own nature, without relation to anything external, remains always similar and immovable.

According to Einstein's views, however, space and time are more intimately connected with one another than it was supposed before, and, within certain limits, the notion of space may be substituted by the notion of time and vice versa. To make this statement more clear, let us consider a railroad passenger having his meal in the dining car. The waiter serving him will know that the passenger ate his soup, steak, and dessert in the same place—i.e., at the same table in the car—without respect to geographical location. But, from the point of view of witnesses at various locations along the track, the same passenger consumed the three courses at different geographical locations, separated by many miles (Figure 16-8*a*, *b*, and *c*). Thus we can make the following trivial statement: **Events taking place in the same place but at different times in a moving system will be considered by a ground observer as taking place at different places.**

Now, following Einstein's idea concerning the reciprocity of space and time, let us replace in the above statement the word "place" with the word "time" and vice versa. The statement will now read: **Events taking place at the same time but in different places in a moving system will be considered by a ground observer as taking place at different times.**

This statement is far from being trivial and means that if, for example, two passengers at the far ends of the diner have their after-dinner cigars lighted simultaneously from the point of view of the dining-car steward, the person standing on the ground will insist that the two cigars were lighted at different times (Figure 16-8*d* and *e*). Since, according to the principle of relativity, neither of the two reference systems should be preferred to the other (the train moves relative to the ground or the ground moves relative to the train) we do not have any reason to take the steward's impression as being true and the ground observer's impression as being wrong, or vice versa.

Figure 16-8. Events that occur at the same place relative to a moving railroad car will look as if they occurred at different places when observed from the ground (a, b, and c). Events that occur simultaneously for an observer in a moving car will look as though they occurred at different times to an observer on the ground (d and e).

Why then do we consider the transformation of the time interval (between the soup and the dessert) into the space interval (the distance along the track) as quite natural and the transformation of the space interval (the distance between the two passengers having their cigars lit) into the time interval (between these two events as observed from the track) as paradoxical and very unusual? The reason lies in the fact that in our everyday life we are accustomed to velocities that lie in the lowest brackets of all the physically possible velocities extending from zero to the velocity of light. A race horse can hardly do better than about one-millionth of a percent of this upper limit of all possible velocities, while a modern supersonic jet plane makes, at best, 0.0003 percent of it. In comparing space and time intervals (i.e., distance and durations), it is rational to choose the units in which the limiting velocity of light is taken to be one. Thus, if we choose a "year" as the unit of duration, the corresponding unit of length will be a light-year or 10,000,000,000,000 km, while if we choose a kilometer as the unit of length, the unit of time will be 0.000003 sec, which is the time interval necessary for light to cover the distance of 1 km. We notice that whenever we choose one unit in a "reasonable" way (a "year" or a "kilometer"), the other unit comes out either too large (a light-year) or too short (3 microseconds) from the point of view of our everyday experience. So, in the case of the passenger eating his dinner on the train, a half-hour interval between the soup and the dessert could result in 200 million miles of distance along the track (time $\times$ $c$) if the train were moving at a speed close to that of light, and we are not surprised that the actual difference is only 20 or 30 miles. On the other hand, the distance of, let us say, 30 meters between two passengers lighting their cigars at opposite ends of the railroad car translates into a time interval of only one hundred-millionth of a second (distance $\div$ $c$), and there is no wonder that this is not apparent to our senses.

The transformation of time intervals into space intervals and vice versa can be given a simple geometrical interpretation as was first done by the German mathematician, H. Minkowski, one of the early followers of Einstein's revolutionary ideas. Minkowski proposed that time or duration be considered as the fourth dimension supplementing the three spatial dimensions and that the transformation from one system of reference to another be considered as a rotation of coordinate systems in this four-dimensional space. His basic idea can be understood by considering the diagram shown in Figure 16-9. In the old system (an observer in the railroad car), the space interval (soup to dessert) and the time interval (1st cigar to 2nd cigar) are both zero. In the rotated coordinate system (corresponding to a moving observer) it is not so, and the two cigar-lightings become non-coincident in time. We notice from this diagram that the appearance of a time interval between two events which were simultaneous in the first system of reference is connected with a shortening of the apparent distance between them as seen from the second system of reference, and, vice versa, the appearance of a space interval between two events which were occurring in the same place of the first system shortens the apparent time interval between them as observed from the second. The first fact gives the correct interpretation of the apparent Fitzgerald's Contraction of the moving bodies, while the second makes the time in a moving system flow slower from the point of view of the second system. Of course, both effects are relative, and each of the two observers moving with respect to one another will see the other fellow as somewhat flattened in the direction of his motion and will consider his watch to be slow.

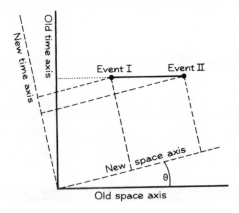

*Figure 16-9. Two events that occur at the same time in the old space-time system occur at different times when the space-time axes are rotated.*

# Advances in Particle Physics

Robert Oerter

## *Inside the Proton*

In 1961, George Gamow published *The Atom and its Nucleus*, which listed 12 "elementary particles." Four years later, the second edition of his *Matter, Earth, and Sky* was published; by then, the list had expanded to 28 particles. "The number of new particles is constantly growing," he noted. "In spite of a very large number of publications bristling with mathematical formulae, very little is yet understood about the nature of elementary particles and the reasons why they behave the way they do." That all changed with the discovery of quarks.

Like the pion, quarks were proposed theoretically before they were discovered experimentally. The basic quark model was developed independently by Murray Gell-Mann and George Zweig in 1963, and it was later extended (by Gell-Mann, Harald Fritsch, and William Bardeen) to include the interactions among the quarks. In this new understanding, most of the particles on Gamow's list are not considered elementary, but are composed of several constituent particles: the quarks.

It is now known that there are six different varieties, or *flavors*, of quarks, as shown in Table A-1.

TABLE A-1  THE FUNDAMENTAL QUARKS

|        | Family |        |
|--------|--------|--------|
| *1*    | *2*    | *3*    |
| up     | charm  | top    |
| down   | strange| bottom |

The six quark flavors are grouped into three families, as shown in the columns of Table A-1. Only the quarks of the first family, the least massive quarks, appear in the ordinary objects around us. Each proton is composed of two up quarks and one down quark, and each neutron is composed of two down quarks and one up quark. The electron is not composed of quarks at all; it remains an elementary particle in the new theory. The other, heavier quark flavors show up only in high-energy particle collisions, such as those that produce cosmic rays or in particle accelerators that are specially designed to produce them. The heavier quarks decay rapidly into lighter flavors, and it is by extreme experimental cleverness only that physicists have been able to deduce their existence.

All of the particles detected in experiments have electric charges that are whole-number multiples of the proton charge. The quarks, however, have *fractional* electric charges. The quarks in the top row of Table A-1 have +2/3 of the proton's charge, whereas those in the bottom row have −1/3 of the proton's charge. (You can easily check that these assignments yield the correct electric charges for the proton and neutron.) Why, then, have fractional charges never been found in experiments? The answer is part of another puzzle. That is, you must ask, "What is the force that holds the quarks together?"

Take a proton, for instance: Since it is very small, the three quarks must be very close together, so that the electric repulsion between the up quarks is very large. There must be another, stronger force that holds the quarks together. This new force is called the *color force* or simply the *strong force*. Here's how it works: Each quark flavor comes in three different colors, which are called red, blue, and green. (The terms *flavor* and *color* are arbitrary labels for these particle properties; they have nothing to do with ordinary flavor and color.) Color can be thought of as a new kind of charge, analogous to electric charge. A positive electric charge can combine with an equal amount of negative charge to produce a neutral object; for example, in a hydrogen atom. In a similar manner, the red, blue, and green color charges can combine to make a color-neutral object (just as red, blue, and green light combine to make white light). According to the quark model, only color-neutral combinations can occur in nature as free particles. A proton, for instance, might be made up of a red up quark, a blue up quark, and a green down quark. Another way to get a color-neutral combination is to combine a quark and an antiquark. The antiquarks come in the opposite colors: antired, antiblue, and antigreen. A pion, for example, might be made of a green up quark and an antigreen down quark. Thus, a single quark can never appear as a free particle. In fact, any color-neutral combination always has a whole-number multiple of the proton's electric charge.

In elementary particle physics, the electric force is the result of a particle's interaction with photons. In the quark model, the color force is the result of the quark's interaction with a new particle, called the *gluon*, because it glues the quarks together. Gluons are electrically neutral, but they carry a color charge. This means that the gluons themselves are subject to the color force, a fact that makes the color force behave in a fundamentally different way than the electric force.

Because of the gluons' self-interaction, the effective strength of the color force gets weaker as you move the quarks closer together and stronger as you move them farther apart—just the opposite of what happens with the electric force! This helps us understand why you never see a free particle with color: The more you try to separate two quarks, the stronger the color force becomes.

A lone quark can never be detected, so all of the evidence for quarks is necessarily indirect. Almost all of the several hundred known particles can be explained as quark-antiquark pairs (the pion, for instance) or three-quark (or three-antiquark) combinations (for example, the proton and the neutron). In addition, modern particle accelerators, colliding electrons and protons at high energy, have probed the internal structure of the proton and revealed evidence of the point-like particles inside. Finally, some high-energy experiments produce particle jets, which are bunches of particles that move in the same direction. These jets are the expected result when a quark inside a proton is given enough of a jolt to break the proton apart.

The theory of quarks, gluons, and their interactions by way of the color force is known as *quantum chromodynamics* or QCD. This theory provides the basis for explaining most of the particles in Gamow's Table A-1 and many of the nuclear reactions discussed earlier. However, some of the particles (electrons, muons, and neutrinos, for instance) and some of the nuclear processes (such as beta decay) cannot be explained in terms of quarks. For these, a whole new set of ideas had to be developed.

## The Weak Force

In Chapter 10, the reader learned of two types of mesons: pions and muons. Both are about 200 times heavier than an electron. The pion is the heavier of the two. Pions play the role that Yukawa predicted for them; they are the particles whose exchange holds the nucleus together. We now know that pions are built of one (up or down) quark and one (up or down) antiquark.

Muons, however, are absorbed in matter much less readily than pions. From this fact, we deduce that their interaction with the nucleus must be much weaker than that of the pion. Physicists sensibly termed this interaction the *weak force*, contrasting it with the strong force by which pions interact with nuclei. Because muons lack the strong interaction, they must not be made of quarks. In modern particle theory, muons are considered truly elementary particles. They have the same charge and spin as an electron, but a larger mass. A third particle, called the tau, also has the same charge and spin as the electron and a larger mass (about 3,500 times the electron's mass), and it interacts through the weak force, not the strong force. The muon and the tau can therefore be considered heavier versions of the electron. Each of these particles is associated with its own type of neutrino, which is produced or absorbed whenever the corresponding particle interacts through the weak force. Like the quarks, these particles can be arranged in three families of increasing mass, as shown in Table A-2.

### TABLE A-2  THE FUNDAMENTAL LEPTONS

| Family | | |
| --- | --- | --- |
| *1* | *2* | *3* |
| Electron neutrino | Mu neutrino | Tau neutrino |
| Electron | Muon | Tau |

Collectively, these six particles (and their antiparticles) are called the *leptons*. No one knows why there are precisely three families of both quarks and leptons. It is appealing to think that the reason might lie in some deep symmetry of nature. There have been many proposals for what a theory with such a symmetry might look like, but as of 2005, none of these has been confirmed experimentally.

Neutrinos (of any family) do not have electric charge or color charge; therefore, they can interact *only* through the weak force. Any process involving neutrinos, then, must proceed by way of the weak force. One important example is the beta decay of radioactive elements. (Alpha decay, on the other hand, occurs by way of the strong force.) The weak force also plays an important role in the thermonuclear fusion reactions that power the sun. These reactions necessarily produce a large number of neutrinos. Because neutrinos interact so weakly with matter, most of those produced in the sun's core travel easily through the sun's outer layers. The earth receives a constant stream of these solar neutrinos, just as we receive a constant stream of photons from the sun. Most of the solar neutrinos pass entirely through the earth without any interactions. About ten billion neutrinos pass through your body every second, without causing you any harm.

In the 1960s, Raymond Davis, a physicist at Brookhaven National Laboratory, decided to build an experiment to detect the solar neutrinos. How does one detect a particle that is capable of passing through the entire earth without any interactions? Even though the probability for any one neutrino to interact with the detector is miniscule, there are so many neutrinos that a large enough detector should be capable of trapping a few of them. Davis used a 100,000-gallon tank filled with a chlorine-containing dry cleaning fluid. A chlorine atom can absorb a neutrino through the weak force and turn into an argon atom, according to the following reaction:

$$Cl^{37} + \nu \rightarrow e^- + Ar^{37}$$

Davis and his collaborators were able to recover about a dozen argon atoms every few months, proving that the expected reaction did indeed take place. The number of argon atoms detected, however, was only half the expected number.

More recent experiments have refined Davis's surprising results and confirmed the solar neutrino deficit. Although the phenomenon remains somewhat mysterious, current theories propose that the electron neutrinos produced in the sun's core are somehow converted into mu neutrinos or tau neutrinos before they get to earth. This can happen only if the neutrinos are not precisely massless. Many experiments are under way to investigate these strange neutrino properties.

## The Standard Model

Since Gamow wrote, tremendous advances have been made in understanding the fundamental structure of matter. His confusing array of particles has been replaced by just six quarks (in three colors each) and six leptons, all of which can be organized into three families, together with a handful of particles that mediate the forces. Not only have we discovered a deeper level of structure, but the strong, weak, and electromagnetic forces have been combined into a unifying theory, which is known as the Standard Model of Elementary Particles. Of the known interactions of matter, only the force of gravity remains outside this framework.

The strong, weak, and electromagnetic forces can all be described using Gamow's picture of two dogs locked in struggle over a bone. In Yukawa's theory of nuclear forces, the bone being exchanged is the pion. In our modern understanding of the strong force, the gluon is the particle that is exchanged between quarks and binds them together. In a similar manner, electromagnetic forces are mediated by the exchange of photons.

For the weak force, we need to introduce three new mediating particles: the $W^+$, $W^-$, and $Z^0$. The Standard Model, developed independently by Abdus Salam and Steven Weinberg in 1967, not only predicted the existence of these three particles, but allowed physicists to predict their masses and decay modes. A race ensued among experimental groups to be the first to detect (or rule out) these new particles. A group led by Carlo Rubbia, an Italian physicist working at the CERN laboratory in Geneva, won the race in 1983, confirming the existence of the $W^+$ and $W^-$, thereby establishing the Standard Model as the basis for understanding particle interactions. A year later the same group bagged the $Z^0$; like the W particles, it had exactly the predicted properties.

The Standard Model predicts the existence of one more particle called the Higgs particle, which has not yet been detected in any experiment. The theory gives few hints about its mass, but there is good reason to believe that it will be discovered when the upgraded CERN accelerator begins operating in 2007.

The Standard Model leaves us with a more orderly list of elementary particles than Gamow's list. The quarks provide the answer to his question of the structure of the particles. Other questions remain unanswered. Are there deeper levels of structure still to be discovered? Is there a reason for the strange pattern of masses of the quarks and leptons? The Standard Model raises new questions, too. Why are there three families of quarks and leptons, rather than one or 100? Can gravity somehow be included in the theory, giving us a unified theory of all the fundamental forces?

In the four decades since Gamow's writings, physicists have indulged in much speculation about these questions. Gamow, however, was reporting on well-established physics, things that were known to be true. I follow his lead, and, having brought you up to date on known physics, resist further speculation. We know a lot more than we did in 1965, when the second edition of *Matter, Earth, and Sky* was published. We are a long way from a "theory of everything," however. New solutions open up new questions, and science goes on.

# Notes

Page 35 *the breaking up of water molecules*—Because it is necessary to *provide* energy (in the form of an electric current) to electrolyze water, one might imagine that the process might be run in the other direction to *produce* electricity. This is indeed possible: Reverse electrolysis is the principle behind modern fuel cells.

One popular type of fuel cell is the proton exchange membrane fuel cell (PEMFC). It is constructed of three layers: an anode, a proton exchange membrane, and a cathode. The anode is made of metal that is etched with channels across its face and coated with powdered platinum. Hydrogen flows through the channels, in which the platinum acts as a catalyst to pull the electrons off of the hydrogen, leaving hydrogen ions (protons). The proton exchange membrane acts as an electrolyte; it allows the protons to pass through to the cathode, but repels the electrons. The electrons flow out through the anode to provide electricity. The cathode is cut with channels and covered with the platinum catalyst just like the anode. Normal air is passed through the channels. Oxygen molecules ($O_2$) from the air are split apart by the catalyst. Each oxygen atom can then combine with two protons to make water ($H_2O$), absorbing two electrons in the process. The result is the process:

$$2H_2 + O_2 \rightarrow 2H_2O + \text{energy}$$

The energy is released in the flow of electric current, as shown in Figure N-1.

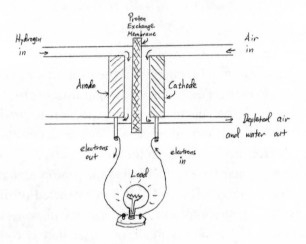

*Figure N-1. The Fuel Cell.*

A fuel cell operates on the same principle as a battery. It uses a sequence of chemical reactions to create a flow of electricity. In normal batteries, however, the chemicals in the anode and cathode are consumed, and after some time, the battery must be discarded or recharged. In a fuel cell, neither the metals nor the catalyst are consumed; only the hydrogen fuel needs to be replenished.

For fuel cells to replace gasoline engines for use in cars depends as much on the production and distribution of hydrogen as it does on the design and manufacture of the fuel cells. Hydrogen is a gas at room temperature and atmospheric pressure, which makes it more difficult to handle than liquid gasoline. The hydrogen must be produced somehow, too; either from other fuels, such as natural gas, or from water by electrolysis. Hydrogen production necessarily involves waste products and some loss of efficiency. As a result, fuel cells provide only a small increase in efficiency over conventional gasoline engines. However, pollution can be more effectively controlled. The fuel cells produce only water as a by-product; the dangerous wastes are confined to the hydrogen production facilities, where they can be dealt with more effectively.

Page 136 *the largest existing cyclotron*—Accelerator technology has progressed more than Gamow could have dreamed in 1965. The size of the world's largest accelerators is now measured, not in inches, but in miles.

Physicists use two main types of particle accelerators today: linear accelerators and synchrotrons. Linear accelerators, as the name implies, accelerate particles in a straight line. The synchrotron is a more sophisticated version of the cyclotron. The particle paths are bent into a circle by a magnetic field, as in the cyclotron. Rather than allowing the particles to spiral outward as they speed up, however, the magnetic field is synchronized with the speed of the particle beam to keep the size of the circle constant.

The most powerful linear accelerator in operation today is the Stanford Linear Accelerator Center (SLAC). It is 2 miles long and accelerates electrons and positrons to 50 billion electron volts. A device similar to the one built by Cockcroft and Walton provides the initial stage of acceleration. For most of the accelerator's length, microwaves provide the acceleration that pushes the particles along like an ocean wave pushing a surfer. In its original mode of operation, SLAC's electron beam was fired into a vat of liquid hydrogen. The resulting collisions probed the inner structure of the proton and provided some of the first evidence for the existence of quarks (see Appendix A for more information). In 1980, SLAC was converted to simultaneously accelerate electrons and positrons and bend the two beams around so that they would collide head-on. This technique raised the total energy to 100 billion electron volts. SLAC experiments helped prove the existence of the charm quark and the $Z^0$ particle.

Until it was shut down in 2000 for an upgrade, CERN's Large Electron Positron (LEP) ring was the world's largest synchrotron. Located near Geneva, LEP is 17 miles in circumference and actually straddles the Swiss-French border. LEP collided electrons and positrons of energies comparable to SLAC's. When the upgrade is completed in 2007, the new machine—renamed the Large Hadron Collider (LHC)—will reach energies of several trillion electron volts. A major goal of the LHC is to look for the mysterious Higgs particle (see Appendix A), predicted to exist by the Standard Model of Elementary Particles, but up until now never detected in any experiment.

Page 181 *the problem of controlled thermonuclear reactions*—Although we have made tremendous progress in understanding the processes of thermonuclear fusion in the past 40 years, Gamow's hope that the problem of controlled thermonuclear fusion would be solved has proven overly optimistic. Even now there is no working power plant producing fusion energy, nor is there any real hope of one in the next 20 years. Research continues, however, and the current pessimistic outlook might prove to be a mistake, just as the optimism of the 1960s. Today's fusion research is pursuing two different paths: *magnetic confinement*, as described by Gamow, and *inertial confinement*, in which a small pellet of fusion fuel is zapped with laser beams from many directions at one time.

For magnetic confinement, the straight-line pinch device Gamow described proved to be unstable. You can picture the difficulty these devices encountered if you imagine squeezing a short spring between your thumb and forefinger. Instead of compressing in a straight line, the spring will tend to bend and bulge outward to one side. The plasma column in the straight-pinch device did exactly the same thing. In physics language, the column was unstable.

Physicists soon discovered the shapes they needed to use to produce a stable plasma. Modern devices that use magnetic confinement are all basically toroidal (shaped like a doughnut). The two main variations are the *tokamak* and the *stellarator*. The tokamak works much like the straight pinch, if you imagine bending the plasma column around into a smooth doughnut shape. One set of electromagnet coils surrounds the doughnut. A second electromagnet rises vertically through the doughnut hole. A changing magnetic field in this second electromagnet induces a current in the plasma. (This is an application of Ampere's electromagnetic induction, which is discussed in Chapter 14) The stellarator, on the other hand, uses an external current in place of the plasma current. This current is produced by winding the electromagnet around the doughnut in a complicated, zig-zag fashion that must be carefully designed. In the stellarator, the tokamak's smooth doughnut shape must be replaced by one that looks half-baked, lumpy, or bulging out on the sides, according to the particular design.

After physicists had solved the stability problem, they faced an even harder task: preventing heat loss from the plasma. It was necessary to develop new diagnostic techniques to measure what was going on in the plasma, and they had to develop new theoretical techniques for predicting and explaining these measurements. The improvement that resulted can be seen in the plasma temperatures reached: 10 million degrees Celsius at a Soviet tokamak called T-3 in 1968 and 300 million degrees Celsius at Princeton's TFTR in 1993. TFTR produced 10 million watts of power for about one second. This is still far from being a usable energy source (it took 40 million watts of external power to accomplish this), still less a commercially viable one, but it was a clear demonstration that thermonuclear fusion can be controlled.

A major international project is currently underway, involving China, the European Union, Japan, Korea, Russia, and the United States. Known as the International Thermonuclear Experimental Reactor (ITER), it calls for a tokamak that is 40 feet in diameter and 12 feet high, and capable of producing 10 times as much power as it consumes. This will be a research device, not a power plant. Most of the energy produced will not be converted into usable forms (such as electricity). Extracting the energy generated by fusion may turn out to be as difficult as producing it. Most of the energy is in the form of fast-moving neutrons, which must be trapped in a "blanket" that surrounds the reactor. Testing various blanket designs is one of ITER's goals.

An alternative approach to controlled fusion is called *inertial confinement*. Rather than attempt to achieve sustained fusion in a large plasma, as in magnetic confinement, this approach seeks a short pulse of energy from a small pellet of deuterium-tritium fuel. The pellet is injected into the exact center of an array of powerful lasers, which fire at one time. The laser energy heats the pellet's outer layer, causing it to implode and creating a hot spot in its center where fusion occurs. The fusion energy explodes the pellet, and the whole process is then repeated. To function as a continuous source of power, the device would need to cycle several times per second.

Inertial confinement, like magnetic confinement, is still far from being a viable energy source. Existing facilities are used for research, not for power generation. The most ambitious project so far is the National Ignition Facility (NIF) at Lawrence Livermore National Laboratory. When completed, NIF will use 192 laser beams to irradiate a fuel pellet two millimeters in diameter. The goal is to release 10 times as much fusion energy as the lasers consume.

In the earlier days of controlled fusion research, the claim was that fusion would be the energy source of the future because it's cheap, clean, and virtually inexhaustible.

Hydrogen is the only thing you need for fuel, and hydrogen is readily available in water. From hydrogen, deuterium and tritium can be produced, perhaps in the blanket of the reactor itself. The main by-product of fusion is ordinary, nonradioactive Helium 4, suitable for inflating party balloons. Compared to the large amounts of radioactive heavy metals produced at a fission nuclear reactor or the tons of greenhouse gases produced by a coal-burning power plant, fusion looks good.

The problem with this rosy picture is that it ignores the effects of the neutrons the fusion process produces. Although crucial for energy generation, the neutron radiation is deadly to humans and makes the entire containment vessel radioactive. The same is true of a fission reactor; however, the inside of a fission reactor is relatively simple. It has none of the electromagnets, exhaust pumps, and blanket components that are needed for a fusion reactor. As soon as the reactor begins operation, all of these components become too radioactive to handle. At ITER, elaborate machinery is being built to repair and replace these systems remotely, using a variety of robotic devices. When the reactor has reached the end of its useful life, the entire containment vessel with the electromagnets, blanket components, and the robotic support machines will become radioactive waste. Even if the multitudinous technical problems are overcome, fusion will never be the cheap, safe, clean energy source it was once advertised to be.

Page 5 *Basic Units Used in Physical Sciences*—Improvements in technology necessitate improvements in the standards of measurement. Precision measurements at the molecular level, for instance, can hardly be based on the distance between two scratches on a metal rod: the scratches might be wider than the distance you want to measure. The definition of the meter and other basic quantities has changed over time to reflect the increasingly stringent requirements of scientists and engineers.

In 1971, the *Systéme International d'Unitès* (SI), more commonly known as the metric system, was established by international agreement. The current definition of the second, according to the SI, is "the duration of 9,192,631,770 periods of the radiation corresponding to the transition between the two hyperfine levels of the ground state of the cesium 133 atom." This definition might not strike the reader as an improvement over the old definition, based on the length of the day. However, the rotation rate of the earth is not constant, so the old definition is unclear. Changes in the flow of the earth's molten interior and external influences, such as the gravitational pull of the moon, combine to cause variations of a few milliseconds in the length of the day. An hour is 1/24 of a day; but *which* day? All cesium 133 atoms, on the other hand, are precisely identical, as far as we can tell. Thus, there can be no variation between the oscillations of two different atoms. The most precise clocks built today are atomic clocks, which keep time by counting the oscillations of (surprise!) cesium 133 atoms. Because atomic clocks are the most accurate way to measure time, it makes sense to use them for the definition of the unit of time.

The meter, too, has been redefined. The new definition is "the length of the path traveled by light in a vacuum during a time interval of 1/299,792,458 of a second." This replaces the older definitions based on the size of the earth, the distance between two scratches on a rod, or the wavelength of light from krypton 86 atoms. Notice that the meter is defined in terms of the unit of time, the second. This definition relies on the fact, known from special relativity, that the speed of light in a vacuum is constant. It is easy to see that the speed of light, according to this definition, is *exactly* 299,792,458 meters per second.

The kilogram is no longer defined in terms of a liter of water; instead, it is now defined as the mass of the standard kilogram, which remains at the International Bureau of Weights and Measures at Sévres. For molecular and atomic masses, scientists use a supplementary standard called the *atomic mass unit* (u). One atomic mass unit is defined as 1/12 of the mass of a carbon 12 atom. This unit, which is based on a physical property rather than a manufactured object, would make a better basis for mass measurements; however, there is not currently a sufficiently accurate method of measuring the masses of everyday objects in terms of atomic mass units.

Page 243 *the first uranium bomb explosion*—David L. Goodstein relates a fascinating story about the first atomic bomb test in *States of Matter* (Englewood Cliffs, NJ: Prentice Hall, 1975, pp. 436–437). Shortly after the test, *Life* magazine published a sequence of photographs of the expanding fireball, giving both the distance scale and the time interval between the pictures. The physicist G. I. Taylor realized that the fireball's expansion rate depended on only the energy released by the blast, $E$, and the density of the surrounding air, $\rho$. The physical dimensions of energy are

$$\frac{|mass||length|^2}{|time|^2}$$

and those of density are

$$\frac{|mass|}{|length|^3}$$

so the relationship between the size of the fireball, $r$, and the time, $t$, must involve the ratio

$$\frac{E}{\rho}$$

because that is the only way to get the dimension of mass to cancel. The dimensions of the combination

$$\frac{E}{\rho} \text{ are } \frac{|length|^5}{|time|^2}.$$

For the size of the fireball, you obviously need something with dimensions of length; the simplest way to obtain such a quantity is to multiply the ratio by $t^2$ and then take the fifth root. This led Taylor to the formula

$$r = \sqrt[5]{A\frac{E}{\rho}t^2},$$

where $A$ is a dimensionless numerical constant. The *Life* photographs gave Taylor several values of $r$ and $t$. The density of air, of course, is known, so the only unknown in the equation is $A$. In physics, dimensionless constants like this are usually around 1, so Taylor was able to use his formula to estimate the value of $E$, the energy released by the bomb. Goodstein remarks, "For the practitioner of the art of dimensional analysis, the nation's deepest secret had been published in *Life* magazine."

Page 266 *the entropy of the entire system cannot decrease*—
The statement, "The entropy of the entire system cannot
decrease," is known as the *second law of thermodynamics*.
There is a great deal of confusion about this law, in part
because those who promote a theological view, rather than
a scientific view, of earth's history have tried to use it as
an argument against evolution. Because this confusion
persists even today (as witnessed by numerous Web sites
that repeat these long-discredited arguments), it seems
worthwhile to expand on Gamow's comment.

In its most naïve form, the argument goes something
like this: "The second law of thermodynamics states that any
system, left to itself, tends to increase disorder. Evolution is
clearly tending toward increased order. Therefore, evolution
contradicts the second law." In this form, the argument is
clearly nonsense. A pond, left to itself, will freeze over
during a cold winter. This involves a change from a highly
disordered liquid state to a highly ordered solid state. The
second law doesn't prevent this decrease in entropy: if it
did, we would have to throw it out because ponds certainly
do freeze over! Rather, the second law requires that the
decrease of entropy of the pond water be compensated
by an increase of entropy elsewhere. In this example, it is
the surrounding environment that absorbs heat from the
pond, with a consequent increase in entropy. As long as
the entropy increase of the environment is larger than the
entropy decrease of the pond water, there is no violation of
the laws of thermodynamics.

A slightly more sophisticated form of the anti-
evolution argument recognizes that the earth is not an
isolated system; it receives energy from the sun. However,

the argument goes on, the sun's energy *increases* disorder. It speeds the processes of breakdown and decay. Therefore, even with an energy source, evolution still violates the second law.

This argument, however, is easily refuted. There is more to the earth's interaction with its "environment" than the energy it receives from the sun. The earth also radiates energy out into space. To properly apply the second law, we have to take into account the change of entropy involved with *both* the absorption and radiation processes. Let's think of the sun as a heat reservoir that maintains a constant temperature $T_1 = 6000$ K. That's the temperature of the radiating surface of the sun, and so it's the effective temperature of the energy we receive from the sun. When the earth absorbs an amount of heat, $Q$, from this reservoir, the reservoir loses entropy:

$$\Delta S_1 = \frac{-Q}{T_1}.$$

On average, the earth's temperature does not increase nor decrease. Therefore, in the same time that it absorbs heat energy $Q$ from the sun's radiation, it must radiate the same amount of heat into space. This energy is radiated at a much lower temperature, approximately equal to the average surface temperature of the earth, $T_2 = 280$ K. We can think of space as a second heat reservoir that absorbs the heat $Q$ and consequently undergoes an entropy increase

$$\Delta S_2 = \frac{+Q}{T_2}.$$

Because $T_1$ is much larger than $T_2$, it is clear that the net entropy of the two reservoirs increases:

$$\Delta S_1 + \Delta S_2 = \frac{-Q}{T_1} + \frac{Q}{T_2} > 0.$$

Even if it is true that the processes of life on earth result in an entropy decrease of the earth, the second law of thermodynamics is not violated unless that decrease is larger than the entropy increase of the two heat reservoirs.

By how much does evolution decrease the earth's entropy? Is it enough to violate the second law? We need to know two things: how fast the entropy of the heat reservoirs increases and how much of an entropy decrease on earth can be reasonably expected. A glance at any astronomy textbook tells us that the earth absorbs $1.1 \times 10^{17}$ Joules of the sun's energy every second, so we can easily calculate the net entropy increase of the two heat reservoirs in one second:

$$\Delta S_1 + \Delta S_2 = \frac{-1.1 \times 10^{17} J}{6000 K} + \frac{1.1 \times 10^{17} J}{280 K} = 3.7 \times 10^{14} \frac{J}{K}$$

It is more difficult to calculate the decrease of earth's entropy due to the processes of life, if, indeed, there is any decrease. None of those who use the entropy argument against evolution even try to estimate it. For comparison, let's calculate something we know how to calculate: the entropy change needed to freeze the earth's oceans solid. The heat energy involved is:

$Q$ = (latent heat of fusion) x (mass of ocean water) =

$$\left( 3.3 \times 10^5 \frac{J}{kg} \right) \times \left( 1.3 \times 10^{21} kg \right) = 4.3 \times 10^{26} J$$

Water freezes at 273 K on the absolute scale, so the corresponding entropy change is:

$$\Delta S_{ocean} = \frac{-Q}{T} = \frac{-4.3 \times 10^{26} J}{273k} = -1.6 \times 10^{24} \frac{J}{K}$$

This is an enormous decrease of entropy. Still, by comparing with the entropy increase of the two heat reservoirs, we see that the second law is violated only if all of the oceans freeze over in a time of $(1.6 \times 10^{24}$ J/K) / (3.7 $\times 10^{14}$ J/K/s) = $4.3 \times 10^9$ seconds or about 140 years.

We need to know one more fact about entropy; that is, it is (roughly) proportional to the mass of the system. The mass of all the living organisms on earth, known as the *biomass*, is considerably less than the mass of the oceans (by a very generous estimate, the biomass is about $10^{16}$ kilograms), so the relevant time frame is actually much shorter. A calculation similar to the one shown previously reveals that the second law of thermodynamics is violated only if the entire biomass is somehow converted from a highly disorganized state (for example, a gas at 10,000 K) to a highly organized state (for example, absolute zero) in about one month. Evolution operates over millions of years; clearly, any decrease of entropy involved is not enough to violate the laws of thermodynamics.

Page 293 *existence of electromagnetic waves*—When James Clerk Maxwell developed his equations for the behavior of electric and magnetic fields, he realized that the equations could describe traveling electromagnetic waves, and he could calculate the speed of the waves from his equations. The wave speed he calculated turned out to be equal to the speed of light! He deduced that light itself must be a form of electromagnetic wave. The wavelength of light was known to be around half a micron ($5 \times 10^{-7}$ meter). The electromagnetic waves generated in Hertz's experiment, by contrast, had wavelengths around a meter—a million times longer than the wavelength of visible light. These are the waves currently used to broadcast video and television signals. In other words, radio waves and visible light are really the same physical phenomenon—electromagnetic waves—they just have different wavelengths.

There are, of course, electromagnetic waves of other wavelengths as well. What is needed to produce them is a means of oscillating electric charge at the appropriate frequency. Microwaves are electromagnetic waves with a shorter wavelength than radio waves. In a microwave oven, the waves are produced by a magnetron, a device that uses a spiraling flow of electrons to generate waves of the desired frequency. Microwaves are also used for communication, such as in cellular phone networks. Wavelengths that are still shorter, but longer than that of visible light, are called *infrared*. The radiant heat that that you feel when you put a hand near a campfire or a hot stove is carried by infrared radiation.

Visible light is produced by the oscillation of charge at the atomic scale, either by the motion of whole atoms (as when the filament of an incandescent light bulb is heated up and glows), or by the movement of electrons between energy levels in individual atoms (as in a fluorescent light). Ultraviolet light is produced in the same way; however, it has a wavelength too short for the human eye to detect. X-rays have even shorter wavelengths; they are produced by firing beams of electrons at a metal target. Still shorter wavelengths, called gamma rays, are produced by the decay of certain radioactive elements and by modern particle accelerators. They are used in medical and security applications and for food sterilization.

Maxwell's discovery meant that the science of optics (see Chapter 6) could be understood in terms of the physics of electric and magnetic fields. He revealed a rainbow of invisible waves, different forms of light with wavelengths too short or too long for the eye to detect. Maxwell's equations allow us to harness these unseen waves for communication, medicine, and industry.

Page 334 *looked at Michelson's failure*—When Gamow states that Einstein "...looked at Michelson's [experiment]... in a much more radical way than did his contemporaries," he implies a much closer connection between the experiment and Einstein's thought than probably existed.

In his definitive scientific biography, *Subtle is the Lord...: The Science and Life of Albert Einstein*, Abraham Pais reviews Einstein's infrequent and somewhat contradictory remarks about the influence of the Michelson experiments. In his original (1905) paper on relativity, Einstein mentioned only "the failed attempts to detect a motion of the earth relative to the 'light-medium;'" he didn't cite any published experimental papers or specify which "attempts" he knew about. In a talk given in Kyoto in 1922, Einstein said, "As a student, I got acquainted with the unaccountable result of the Michelson experiment and then realized intuitively that it might be our incorrect thinking to take into account the motion of the earth relative to the ether.... In effect, this is the first route that led me to what is now called the special principles of relativity…."

In later years, however, Einstein downplayed the influence of Michelson's experiments. At one point, he denied that he knew about them before his 1905 paper, but on further reflection he suggested that they influenced him indirectly through H. A. Lorentz's 1895 paper (which Einstein also mentioned in his Kyoto talk). Pais concludes that Einstein must have known of Michelson's results, but that his (Einstein's) thinking was influenced more strongly by two other experimental results, the aberration of starlight and Fizeau's experiment on the speed of light in moving water. Although the results of both experiments could be explained in terms of the ether hypothesis, Einstein was seemingly dissatisfied with these explanations. He sought a deeper, more comprehensive solution, and he found it, in the special theory of relativity.

# Index

Page numbers followed by *n* signify footnotes.